The GOAT
A Natural and Cultural History

First published in the United States and Canada in 2020 by
Princeton University Press
41 William Street
Princeton, New Jersey 08540
press.princeton.edu

First published in the United Kingdom in 2020 by
Ivy Press
An imprint of The Quarto Group
The Old Brewery, 6 Blundell Street
London N7 9BH, United Kingdom

Library of Congress Control Number: 2019956217

ISBN: 978-0-691-19133-1

This book was conceived, designed, and produced by
Ivy Press
58 West Street, Brighton BN1 2RA,
United Kingdom

Publisher David Breuer
Editorial Director Tom Kitch
Art Director James Lawrence
Project Editor Angela Koo
Design JC Lanaway
Picture Research Sharon Dortenzio, Jane Lanaway, Sue Weaver
Illustrator John Woodcock
Consultant Debbie Cherney

Printed in Singapore

10 9 8 7 6 5 4 3 2 1

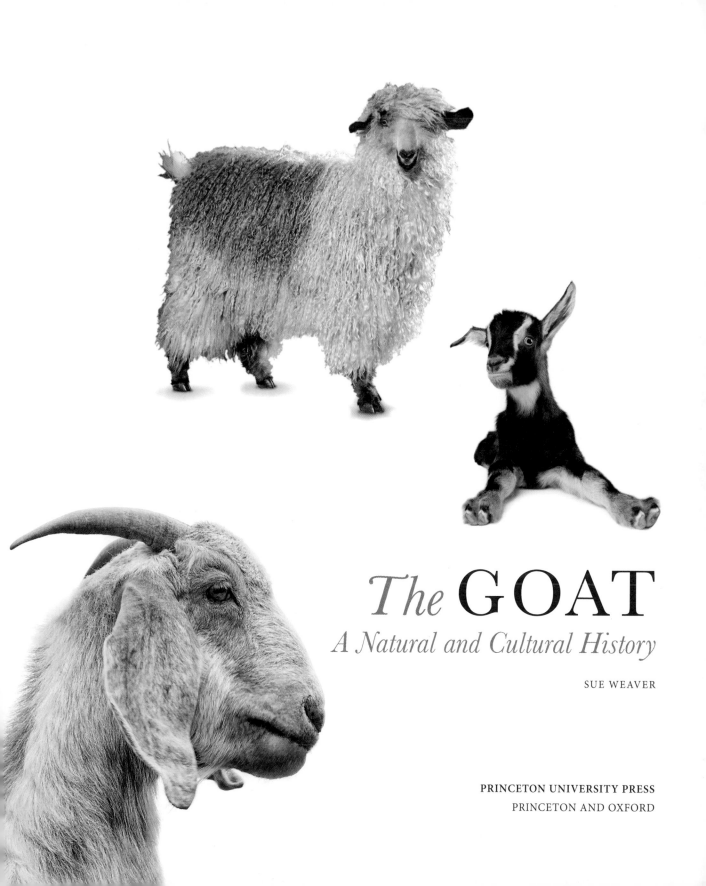

The GOAT
A Natural and Cultural History

SUE WEAVER

PRINCETON UNIVERSITY PRESS
PRINCETON AND OXFORD

Contents ✐

Introducing the Goat   6

CHAPTER 1
Domestication

Goat Taxonomy   14
Goats Come in from the Wild   16
Dispersal Throughout the World   22
Feral Goats   26

CHAPTER 2
Anatomy & Biology

Anatomy of a Goat   30
The Life Cycle of a Goat   32
The Skeleton   34
The Ruminant Digestive System   36
Lips & Teeth   40
The Five Senses   42
Hooves   46
Horns   48
Wattles   52
The Mammary System   54
The Inheritance of Traits   58
Colors & Markings   60
Sheep versus Goats   64
Myotonic Goats   66
Dwarfism in Goats   68

CHAPTER 3
Society & Behavior

The Goat Lifestyle   72
Hierarchy   76
Bucks   80
Does   84
Early Life   86
Vocalization   92
Are Goats Smart?   94

CHAPTER 4
Goat Management

Caring for Hooves	98
Kidding	100
Common Health Issues	104
An Unlikely Predator	108
Goats Behaving Badly	110
Handling & Training Goats	112
Goat Play	116

CHAPTER 5
Goats & People

Goats for Milk	122
Goats for Meat	126
Goats for Fiber	130
Goats for Brush Control	134
Packgoats	136
Harness Goats	138
Goats in Mythology & Folklore	142
The Sacrificial Goat	146
Military Mascots	148
Goats & Musical Instruments	152
Goats in Popular Culture	154
Goat Quotes & Proverbs	158

CHAPTER 6
A Directory of Goat Breeds

How Breeds Evolve	162
The Directory	164

Appendices

Glossary	214
Bibliography	216
Biographies & Acknowledgments	219
Index	220
Picture Credits	224

Introducing the Goat ✍

Although we cannot be entirely sure—and research continues to give new insights into their origins—it is generally believed that goats were domesticated in the Middle East roughly 10,000 years ago. Along with sheep, who were also domesticated at the same time and in the same locale, they were the first domestic livestock species.

Easy to tame, easy to feed, hardy, and docile, goats meant life to the ancients. These animals provided meat and milk; skins and pelts that could be used to fashion water carriers and body coverings, and sinew with which to sew them; bones and horns for tools; soft undercoats for spinning and felting; and hair for crafting tents. In addition, burning goat dung kept primitive shelters warm, and dung could also be used as a fertilizer to enrich the soil. As pack animals, goats carried loads and pulled travois and other forms of sled. They made fine trading goods for people on the move, and trade in turn helped distribute goats around the world. Later in Africa and the Middle East, wealth was reckoned in goats and maintaining huge herds was a major status symbol of the day. Bride prices and dowries were paid in goats, and goats provided sacrifices for the gods.

Even in our age of spit-and-polish show goats and high-producing dairy does, everyday goats provide life-giving milk and meat in developing countries, where no other livestock would thrive. Goats are the perfect livestock for marginal living situations.

Opposite: Docile, intelligent, and easy to tame, goats have been a friend to humankind for thousands of years.

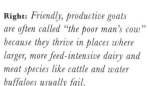

Right: Friendly, productive goats are often called "the poor man's cow" because they thrive in places where larger, more feed-intensive dairy and meat species like cattle and water buffaloes usually fail.

Goats are primarily browsers, not grazers, so they thrive on brush, leaves, and coarse plants that other species leave behind or can't digest. Goats can stand on their hind legs to browse low-hanging tree branches that are up to six or seven feet off the ground. They're strong climbers, too, so they can forage in steep places where sheep and cows won't venture. They can cover great distances looking for food, and will tolerate extremes in temperature and handle heat stress and prolonged water deprivation when they must. As Thomas Bewick observed in *A General History of Quadrupeds*, as far back as 1792, "The goat is an animal easily sustained, and is chiefly therefore the property of those who inhabit wild and uncultivated regions, where it finds an ample supply of food from spontaneous productions of nature, in situations inaccessible to other creatures."

All of these qualities make goats ideal livestock for the poor. They aren't expensive to buy, house, feed, or maintain. And having several goats provides more sales and trading opportunities than keeping a single large cow or water buffalo; several goats take up less space, too. Goats even reproduce without fuss, providing tasty kids for the larder.

Goats are gentle, intelligent, and affectionate, but they are also independent, mischievous, determined, and frustrating. In this book we'll show you where goats came from, how they've interacted with human caretakers through the centuries, and what makes them tick. If you're ready to learn all things goat, read on.

Below left: *Yes, goats in trees! Goats are amazing climbers; they're descended from mountains goats after all. These agile goats are browsing argan tree fruit in Morocco.*

"GOAT" IN ENGLISH

The word "goat" comes from the Old English *gāt* (in turn, from the Proto-Germanic *gaito*), meaning a female goat. A male goat was a *bucca* or *gātbucca*, which evolved into "buck," the correct term until a shift to *he-goat* and *she-goat* occurred in the late twelfth century. "Nanny goat" originated in the eighteenth century, and "billy goat" in the mid-nineteenth century. "Doe" and "buck" then became popular in the mid-twentieth century.

ABOUT THIS BOOK

Our story of the goat begins in Chapter 1 as we take a look at the road goats have traveled through history, from prehistoric beast to man's four-legged helper. We consider true wild goats, among them *Capra aegagrus*, first domesticated in the Fertile Crescent around 10,000 years ago and ancestor to today's goats worldwide, and why it was relatively easy to bring him in from the wild. We also see how goats spread quickly around the globe in the company of Neolithic herders and, later, colonists, seafaring traders, and explorers, sometimes escaping their handlers and returning to the wild as feral goats.

Chapter 2 then addresses biology and anatomy. We survey goats' bones, their organs, their hooves, their splendid horns, and each of their five senses. We lightly touch on genetics, on dwarfism, and what makes some goats "faint." We also examine a wide array of colors and markings before taking a peek at the differences between the goat and its cousin, the sheep.

In Chapter 3 we delve into goat society and behavior, and what makes goats tick. This chapter explains how they interact with one another and establish a hierarchy, and why this is important to goats. We consider courtship and the goat—when and how they breed, and how they choose their mates. We look at how their early lives unfold and how kids play and mature. We also consider vocalizations—why goats call and what it means. Are goats smart? They are, and we'll show you why we know it's so.

In Chapter 4 we look at properly caring for goats: trimming their hooves, attending them as they give birth, and addressing common health issues. We consider the hows and whys of handling and training them, and how to troubleshoot behavior quirks.

Above: *Feral goats like these kids descend from domestic goats gone wild. Feral goats still exist and thrive in parts of Britain, the US, New Zealand, and Australia.*

We round things out in Chapter 5 by observing how goats and humans interact: milk goats, meat goats, fiber goats, goats for vegetation control, even packgoats that carry your gear when you go camping. We learn that goats were once widely used as cart and carriage goats, and how they still continue to perform this function today. Along the way we'll consider the goats that used to pose with children in vintage photos, and hauled the wagons of eccentric "goat men" in the early twentieth century. Goats abound in world myth, too, and we'll take a peek at some of them: the goats who pulled Thor's chariot, the Yule goat of Scandinavia, and many goat deities and fairies from around the world. We'll consider goats raised to become religious sacrifices, past and present, and military mascots; we'll even explore the use of goatskins to make musical instruments, from drums to bagpipes, horns, stringed instruments, and beyond. This chapter concludes with a look at goats in popular culture.

Finally, Chapter 6 takes the form of a breed directory at the end of the book, introducing a gallery of 48 glorious goat breeds from around the world, with key facts and figures on each one. Read on and meet the wondrous goat—you may never look at these animals in the same way again.

Above: *Goats were once used to pack goods in mountainous areas like Mongolia and Nepal. Now they pack camping gear for walkers and backwoods adventurers in North America and Europe.*

10 FACTS ABOUT GOATS

1. According to the United Nations Food and Agriculture Organization, the country with the largest goat population is China, followed by Pakistan, and then Nigeria.

2. India is the world's largest producer of goat milk, followed by Bangladesh, and then France. Most of India's and Bangladesh's production is used for home consumption, while roughly 90 percent of France's milk is used for crafting cheese.

3. Goat meat is eaten throughout the world, and kid is a special delicacy. In the Caribbean and some parts of Asia, including India, Pakistan, Bangladesh, Sri Lanka, and Nepal, English speakers call it "mutton," whereas in the UK, US, and Australia, mutton refers to the meat of adult sheep.

4. Of 570 goat breeds found worldwide, only 69 are specialized dairy breeds. And 221 breeds originated in Europe and the Caucasus, followed by Asia with 183, and Africa with 96. North America has the fewest with 6.

5. The life expectancy for a well-cared-for goat is around 12 to 15 years of age. According to the Guinness World Records, the oldest known was an English doe named McGinty that lived to 22 years and 5 months.

6. Cartoons and folklore notwithstanding, goats don't eat tin cans, garbage, or clothing. While they can and often do subsist on meager fare, goats are in fact fastidious eaters—for instance, they won't eat feed that is soiled with feces or has fallen on the ground—when provided with a well-balanced diet.

7. Unlike cartoon "billy goats," unaltered male goats rarely rush people when their backs are turned and bash them with their horns. Aggressive goats are more likely to face whomever they're threatening; they will lower their heads, with horns jutting forward, or they will rear on their hind legs and swoop down toward the person they're intimidating. Even this is rare because, unless they've been mistreated, most male goats are as friendly as females and wethered (castrated) males.

8. Goats are curious, sociable, and intelligent. According to a study carried out by researchers at London's Queen Mary University in the UK, they are able to communicate and interract with humans to the same degree as domestic dogs. (For more on these findings, see Chapter 3.)

9. Most goats are born with horn buds that, unless growth is artificially inhibited (as described in Chapter 3), develop into horns. The longest horns on a living goat in 2018, according to Guinness World Records, belonged to Rasputin, a Valais Blackneck buck from Austria, whose horns measured an impressive 55 in. (140 cm) from tip to tip.

10. Despite popular thought, goats aren't stinky. The only smelly goats are bucks in rut (explained further in Chapter 3). During breeding season, glands near their horns secrete strong-scented, greasy musk, and they squirt thin streams of urine onto themselves. Does and wethers don't do this, so unless they're in close contact with a rutting buck, goats kept in clean surroundings don't stink.

CHAPTER 1

Domestication

Goat Taxonomy ❧

Goats are members of the Artiodactyla order, the Bovidae family, the subfamily Caprinae, and the genus *Capra*. Fossil and molecular remains suggest that the common ancestor of today's goats originated in an alpine region of Asia or the Mediterranean during the late Miocene 11 million years ago (mya). The fossil remains of a proto-goat called *Pachytragus crassicornus* that lived roughly 8.2 million years ago were unearthed at Pikermi in Greece and Maragheh in Iran. And in 2006, bones of the first true goat (*Capra dalii*), a creature that lived 1.77 million years ago, were found near Dmanisi in Georgia. This site was once believed to be the oldest human settlement outside of Africa; the fact that *Capra dalii* bones were found alongside those of *Homo erectus georgicus* indicate that it was being hunted by this time. But unfortunately, very little is known about how goats evolved from *Capra dalii* into the various wild species we know today.

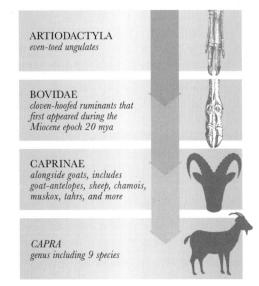

ARTIODACTYLA
even-toed ungulates

BOVIDAE
cloven-hoofed ruminants that first appeared during the Miocene epoch 20 mya

CAPRINAE
alongside goats, includes goat-antelopes, sheep, chamois, muskox, tahrs, and more

CAPRA
genus including 9 species

Left: *The biological classification of the* Capra *genus. The genus in turn consists of nine species, as shown opposite.*

Below: *This old German engraving depicts* Capra ibex *(left) and* Capra falconeri *(right), two species widely hunted throughout their range long before goats were domesticated.*

CAPRA SPECIES

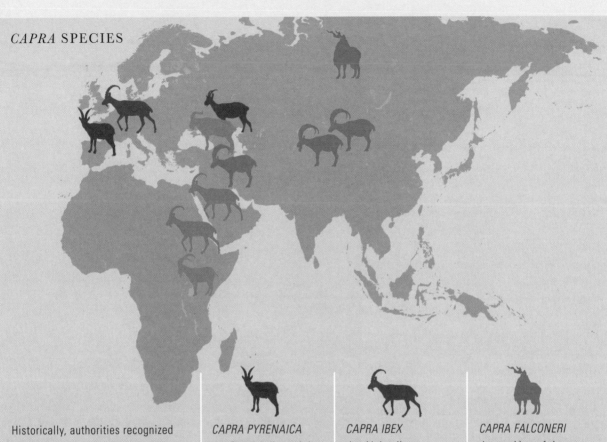

Historically, authorities recognized only two caprine species: markhors (*Capra falconeri*) and everything else. Later, classifications changed. Initially, Spanish (*Capra pyrenaica*), Siberian (*Capra sibirica*), and walia (*Capra walie*) ibexes were considered subspecies of the Alpine ibex (*Capra ibex*). Now they're considered to be separate species, bringing the modern count to nine. The domestic goat (*Capra hircus* or *Capra aegagrus hircus*), found everywhere around the globe except Antarctica, is a subspecies of the bezoar (*Capra aegagrus* or *Capra aegagrus aegagrus*). Today's species are shown to the right.

CAPRA PYRENAICA
the Spanish ibex of the Iberian Peninsula

CAPRA IBEX
the Alpine ibex of the Alps

CAPRA FALCONERI
the markhor of the western Himalayas

CAPRA AEGAGRUS
the bezoar ibex of the Caucusus, Central Asia, and the Middle East

CAPRA SIBIRICA
the Siberian ibex of Central Asia

CAPRA NUBIANA
the Nubian ibex of northeastern Africa and parts of Arabia

CAPRA CAUCASICA
the West Caucasian tur of the West Caucasus

CAPRA CYLINDRICORNIS
the East Caucasian tur of the East Caucasus

CAPRA WALIE
the walia ibex of the northern Ethiopian mountains

Goats Come in from the Wild ✐

Goats figured prominently in Neanderthal cave paintings in both France and Spain, dating at least as far back as 64,000 years ago and possibly even longer—roughly 20,000 years before modern humans (*Homo sapiens*) arrived in those regions. The first interactions between the earliest *Capra* and *Homo* species would have been as hunter and prey. Goats were clearly a favored species for Neanderthal hunters, as evidenced by a treasure trove of Caucasian tur bones excavated at Ortvale Klde, a rock shelter in the South Caucasus, Georgia, that dates to between 60,000 and 20,000 years ago, in addition to bones found in ancient food middens throughout the Fertile Crescent region of the Middle East (see map opposite).

While hunting fulfilled a need, game animals needed to be pursued when and if they were available, so it would eventually prove far more advantageous to keep them close at hand. Early on, therefore, man experimented with domesticating various species to ensure a steady supply of meat and hides. Some animals, such as gazelles, resisted domestication. Others, like goats, dogs, and sheep, accepted man's efforts exceedingly well.

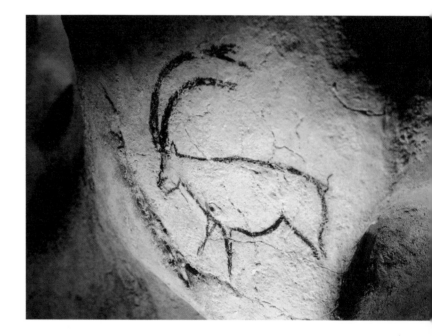

DOMESTICATING GOATS

The Fertile Crescent was a domestication center for a wide array of plant and livestock species, including lentils, barley, rye, flax, chickpeas, cows, pigs, sheep, and goats. This region is generally believed to be the oldest known site of goat domestication, dating back to roughly 10,000 years ago, although where and when this occurred is still subject to debate (see page 20).

Above: *A drawing of a goat found on the walls of the Chauvet Cave, in France's Ardèche region, dating back 36,000 years. (The image shown here is a replica of the original drawing.)*

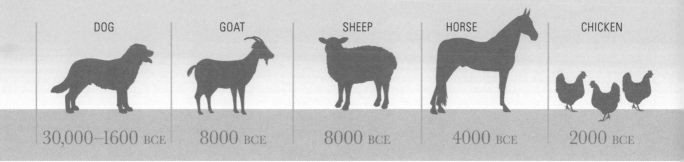

DOG	GOAT	SHEEP	HORSE	CHICKEN
30,000–1600 BCE	8000 BCE	8000 BCE	4000 BCE	2000 BCE

Above: Animal domestication dates are subject to much debate, but goats and sheep are believed to have emerged in the same region, and at around the same time.

Below right: *Initially, scientists believed that early humans domesticated more than one species of wild goat. However, DNA testing indicates only* Capra aegagrus, *the bezoar ibex, was involved.*

Certain species were ripe for domestication because they typically behaved in certain ways. For instance, they had to be able to fend for themselves, finding enough to eat around human settlements to survive without having to compete with humans for valuable food resources. They had to reach productive maturity quickly, whether for slaughter, draft, or breeding purposes. They had to be willing to breed in captivity and not be terribly picky about their mates. They needed to bond quickly with their

offspring and their young had to be mobile soon after birth. They also had to be docile by nature—not nervous, reactive, or prone to panicking—and amenable to living in enclosures with others of their kind. Finally, they had to conform to a strong social hierarchy in which humans could assume leadership. Goats would have shone in every way, which is why they were, along with sheep, probably one of the very first livestock species to be domesticated.

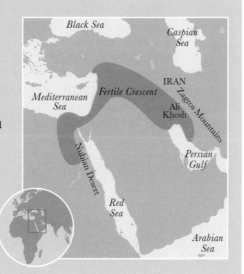

THE FERTILE CRESCENT

Research has shown that goats were first domesticated in the Fertile Crescent, the "cradle of civilization" in the Middle East. The area was comprised of modern-day Iraq, Syria, Israel, Egypt, Palestine, Lebanon, Jordan, and parts of Turkey and Iran.

Black Sea

Caspian Sea

IRAN

Mediterranean Sea

Fertile Crescent

Zagros Mountains

Ali Khosh

Nubian Desert

Persian Gulf

Red Sea

Arabian Sea

Goat bones found in food middens in the region had mostly been those of large old bucks—hunters would have targeted those animals that provided the most meat. But in the 1960s, the Neolithic settlement of Ganj Dareh was discovered in the Zagros Mountains of Iran. The goat bones unearthed here largely came from immature bucks, indicating that residents had kept does and a few mature bucks for breeding, and had slaughtered and eaten young bucks. During Ganj Dareh's 100 to 200-year-long occupation, these newly domesticated goats still looked like their bezoar ancestors (see page 21) and the wild bezoar goats nearby; in other words, domestication had not progressed enough to create physical changes (see box).

Very early domestic goat bones were also found at Ali Kosh, a semi-arid plain located south of Ganj Dareh and settled about 500 to 1,000 years later. The bones found at Ali Kosh clearly indicated domestication changes; these goats had become significantly smaller, and the bucks' horns had become smaller, too, while most does had no horns at all. Since Ali Kosh is located outside of bezoar goat range, it's likely that the ancestors of this settlement's goats had originated at Ganj Dareh.

Opposite: Whether or not it's finally determined that domestication initially occurred at Ganj Dareh, it's almost certain that it occurred in the Zagros Mountains of modern-day Iran, Iraq, and southern Turkey, where signs of agriculture date back to 9000 BCE. Sheep were domesticated in the region at the same time as goats.

DOMESTICATION SYNDROME

When a species becomes domesticated, physical changes gradually occur that set them apart from their wild ancestors. Over time, goats became increasingly tamer and their horns became much smaller. Coats also grew longer, more luxurious, and sometimes wavy—or came to resemble the short, slick coat of a seal; they also developed white spots on their face and bodies, and, eventually, breed-specific coat markings. Many developed longer, floppier ears and shorter muzzles. Their cranial capacity and brains diminished in size, as did their teeth, and estrus cycles became more frequent. And through selective breeding, some grew smaller, while others became much larger. Today's diminutive Pygmy goats (see pages 68–69) represent the lower end of the scale, whereas some of the massive Pakistani and Indian breeds—including Beetals (left) and Kamoris—now break world records, with Beetals sometimes topping 660 lb. (300 kg).

A NEW LOOK AT DOMESTICATION

While Ganj Dareh has generally been considered the site of the first domestication of goats, many experts have also proposed that primary domestication occurred elsewhere in the world, too, such as in China, India, and Pakistan. However, DNA sequencing is now starting to fill in the gaps.

Unlike nuclear DNA, which is inherited from both parents (see pages 58–59), mitochrondrial DNA (mtDNA) is passed exclusively from a mother to her offspring, and so can be used to study matrilineal lineage. Analysis of mtDNA taken from a large number of modern goat breeds from around the world has established that today's goats evolved through six distinct matrilineal lines, with the members of each of these six groups—"haplogroups"—sharing a common ancestor.

Some believe that haplogroup A descends from a bezoar doe from the Zagros Mountains (probably Ganj Dareh), but there are others who suggest that this haplogroup originated in Eastern Anatolia, where it's common in today's wild bezoar populations. Recent archaeological evidence also points to domestication in this region, beginning around 11,000 to 10,500 years ago, which would make it earlier than the Ganj Dareh settlement. Domestic goat bones from Anatolia also suggest that physical changes such as reduced size and smaller horns were taking place by this point. As for the smaller haplogroups, the indication is that domestication in each case can be traced back to a goat native to that particular location. With further study we will know for certain.

GOAT HAPLOGROUPS

In 2018, data for 83 wild and domestic pre-Stone Age to medieval goats from sites around the Fertile Crescent was compared to ancient and modern samples from Africa, Europe, Asia, and the Middle East. Most of the domesticated ancient goats belonged to mitochondrial haplogroups that still exist in modern goats.

Haplogroup	Country or region	Occurs in % of modern goats
A	Worldwide	90.86%
*B1	Asia, sub-Saharan Africa, and Greece	4.4%
*B2	China and Mongolia	1.44%
C	Asia and Europe	1.44%
D	Asia and Central Europe	0.54%
F	Sicily	0.12%
G	Near East and Northern Africa	1.11%

*Haplogroup B is divided into two subgroups, B1 and B2, which initially shared a common ancestor then diverged.

Above: *Wild* Capra aegagrus, *depicted in this engraving from 1836 and ancestor of today's domestic goats, is still found in the mountains of Asia Minor and the Middle East.*

THE BEZOAR CAME FIRST

So although some authorities still believe that the markhor influenced a number of today's Asian goat breeds, it is therefore very likely that the only ancestor of today's goats was in fact the bezoar goat—also known as the bezoar ibex, Persian ibex, pasang (Farsi for "rock-footed"), and *Capra aegagrus* (from the Greek *aigagros*, meaning "wild goat"). The bezoar today is a critically endangered wild goat found in small numbers in the mountains of Asia Minor and the Middle East. It's also found on some Aegean Islands and on Crete, where it's thought that the animals represent relic populations of early domesticated goats that were taken to the Mediterranean islands during the Neolithic Revolution and that now live in feral populations. (For more on the bezoar, see page 164.)

WILD, TAME, OR DOMESTICATED?

Humans undoubtedly tamed goats prior to their domestication in the Middle East. Individual wild goats would have been fairly easily tamed if captured as kids and raised with kindness. But domestication is more than taming— it occurs when people selectively breed a species to mold it into something beneficial to humans, and the process takes generations. True wild goats are not only wild, but they descend in full from one of the nine wild species of the genus *Capra*.

Dispersal Throughout the World ◈

Whether goats were first domesticated at Ganj Dareh, in Eastern Anatolia, or even a heretofore undiscovered domestic center, they spread fairly quickly throughout the Middle East, probably through trade and possibly through resettlement, as evidenced by significant archaeological finds.

NEW TERRITORIES

Over time, climate changes and population growth inspired Neolithic farmers to migrate yet farther. Domestic animals provided a movable food supply to make this possible, and goats were especially useful as they would follow

Above: *This small bronze figure of an ibex dating to around 1000 BCE is typical of objects found in Early Iron Age burials in western Iran.*

Below: *This drinking vessel, depicting a winged goat, originated in Nisa, Turkmenistan, in the second century BCE, when Nisa was the capital of the Parthian Empire.*

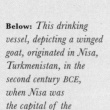

KEY ARCHAEOLOGICAL FINDS

• Aşıklı Höyük, Turkey (occupied 8000–7800 BCE). Excavation indicates that goats and sheep were held captive within the settlement, an early sign of domestication. Some think that this was the Anatolian site where goats were first domesticated.

• 'Ain Ghazal, Jordan (7250–5000 BCE). A large settlement of about 3,000 people. Goats, first wild and then domestic, were a major food source here, where many recovered leg bones suggest tethering.

• Mehrgarh, Pakistan (7000 BCE). Residents raised domestic sheep and goats by 5500 BCE. This is one of the earliest sites with evidence of farming and herding in South Asia, and was once thought to be a domestication site for goats.

• Göytəpə, Azerbaijan (6000–5500 BCE). Although residents also hunted wild game, more than 80 percent of the recovered animal bones here were from domestic goats, sheep, cattle, and pigs.

• Jeitun, southern Turkmenistan (6000–3500 BCE). Clay figurines, decorated ceramics, and small stone axes recovered from Jeitun show similarities with those of Neolithic sites in the Zagros Mountains, including Ganj Dareh. Jeitun's residents kept domestic goats and sheep and also hunted wild goats.

a leader and were easy to transport by land or boat. These migrations followed three different routes.

The first major route—the Danubian—followed valleys along the Danube and Rhine Rivers north and east into the plains of central and northern Europe. A small distance took considerable time to cover, moving on foot and allowing herds to graze or browse. Staying fairly close to rivers and their tributaries, some parties established temporary settlements along the way. Others settled permanently in southern Hungary, Moravia, and Bohemia. Goats, with their herders, reached the Balkans by around 6000 BCE, central Europe by 5400 BCE, and Britain and Scandinavia by roughly 4000 BCE.

A second route took place by sea. Following the northern shorelines of the Mediterranean, these reached Greece and Cyprus by around 7000 BCE, the Italian peninsula by 6000 BCE, southern France by 5700 BCE, and the southern and western coasts of Iberia (present-day Spain) by around 5400 BCE.

A third route followed the Mediterranean Sea along its southern shores and then down into Africa, where it proceeded east along the coast of North Africa, probably reaching Haua Fteah in Libya by 4800 BCE and Grotte Capeletti in the Sahara by 4000 BCE. Goats and their herders then eventually traveled south through Africa between 5,000 to 4,000 years ago, eventually reaching the south of the continent by 20 CE.

Below: *Goats were dispersed throughout Europe and Africa via three routes. In some instances they reached destinations in Eastern Europe as part of multiple dispersals.*

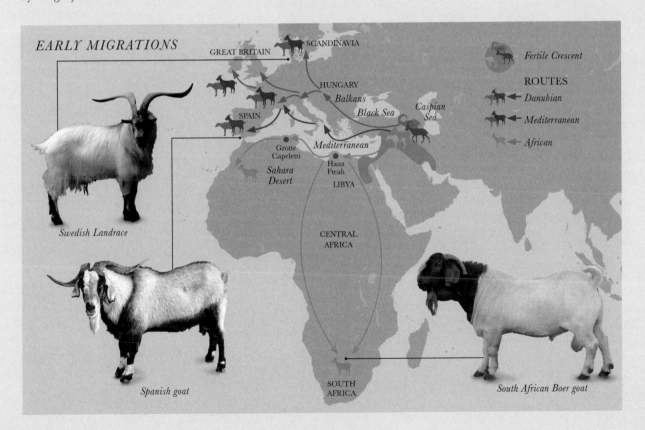

EARLY MIGRATIONS

SCANDINAVIA
GREAT BRITAIN
HUNGARY
Balkans
Black Sea
Caspian Sea
SPAIN
Grotte Capeletti
Mediterranean
Sahara Desert
Haua Fteah
LIBYA
CENTRAL AFRICA
SOUTH AFRICA

Fertile Crescent

ROUTES
Danubian
Mediterranean
African

Swedish Landrace

Spanish goat

South African Boer goat

THE AGE OF EXPLORATION

Although long established in Europe and Asia, goats became truly worldwide once explorers began to discover the New World, starting in the late 1400s, and emigrants began to settle in new lands. Spain's contribution during this era was typical of many nations.

In all, Spain colonized the Philippines, Hispaniola, Puerto Rico, Jamaica, Cuba, Trinidad, half of South America, most of Central America, Mexico, Florida, the Southwestern and the Pacific-Coast regions of the US, and territory in North Africa and Oceania. Wherever they went, Spaniards took goats.

Goats were portable and hardy enough for shipboard life, and could be used to provide meat, milk, hides, and waste disposal en route. In addition, Spanish sailors released goats on uninhabited islands along their nautical routes, knowing that the animals would survive and multiply, and would therefore provide fresh meat on subsequent voyages. Goats also accompanied explorers following land routes as a walking meat supply, escaping, at times, to establish feral populations (see pages 26–27).

Below: *During the age of exploration, goats were carried aboard ships to provide milk and meat at sea as well as for stocking uninhabited islands.*

The famous Scottish castaway Alexander Selkirk was a Royal Navy officer who, after a dispute with his captain, was abandoned on the uninhabited Juan Fernández Islands of Chile. Luckily for Selkirk, other sailors had previously visited the islands and left behind some goats, thereby establishing the beginnings of a feral population. Selkirk managed to survive for more than four years, consuming the meat and milk provided by these goats, and even making clothes from their hides.

Below: *The British explorer Captain James Cook setting out in the HMS Endeavour. Cook carried goats to New Zealand, Australia, and Hawaii, some of which he presented to indigenous chieftains.*

Although most releases went unrecorded, English, Portuguese, and Dutch sailors also salted islands with goats. One of the best-known examples is that of British explorer Captain James Cook, who made three voyages to the Pacific between 1768 and 1779 onboard the HMS *Endeavour*. On their travels, his crew released goats on Niihau in the Hawaiian archipelago, as well as on several islands off the coast of New Zealand. And in 1853, American Commodore Matthew Perry, on the first of his visits to Japan as part of a mission to open up Japanese ports to trade with the West, released six goats in the Bonin Islands at the southern tip of Japan.

Feral Goats

Feral goats are domestic goats that have escaped their handlers or have been purposely released to fend for themselves. Today these goats live life wild and free in spots around the globe as diverse as Australia, New Zealand, Crete, Hawaii, Great Britain, Chile, Mauritius, Japan, and the western United States, where they are often considered problematic. Feral goats annoy landowners by competing with both livestock and native animals for food and water resources. They can also deplete the soil's protective vegetation and crust with their hooves, and can affect trees and shrubs by preventing the regeneration of seedlings. They will also overgraze grass when brush is scarce.

As mentioned previously, goats were popular shipboard animals, and were often purposely released on uninhabited islands to proliferate, which has resulted in them being present on a great many islands throughout the world. However, because predators were largely lacking in these places, their populations skyrocketed, upsetting the fragile local ecology. And it's a problem that persists worldwide today.

Right: *Australia's huge population of feral goats causes major economic and environmental damage through overgrazing and competition with livestock and native wildlife.*

GREAT ORME

Not all feral populations are vilified. Some of the smaller herds across Europe are tolerated and even semi-protected for their historical value and for the impact they have on tourism. One of the best known of these is the herd of feral Kashmiri goats (currently numbering 122) that roam the Great Orme County Park in Llandudno, on the northeast coast of Wales. Two of Queen Victoria's Windsor herd of Kashmiri goats (see also pages 150–51) were acquired in the late 1800s by Sir Savage Mostyn for his estate Gloddaeth Hall, in Llandudno. Then, around 100 years ago, the herd that had descended from this pair were transferred to Great Orme. To keep numbers in check without ongoing culling, some of the does are given contraceptive injections. All of the Great Orme goats are white, with long hair and puffy bangs. The bucks have immense scimitar horns while the does' horns are short and slim.

GOATS DOWN UNDER

Although prolific around the world, nowhere are feral goats quite such a problem as they are in Australia. Nineteen goats came to Australia with the First Fleet in 1788 and more soon followed. Ideally suited to Australia's climate, hardy goats escaped and flourished. Then, between 1860 and 1900, entrepreneurs imported cashmere goats in a misguided effort to establish an Australian fiber industry. When it failed, these goats were released into the wild, and they joined the feral population.

Due to a lack of natural predators except for dingoes, the goat population can increase at a staggering rate. According to Queensland government statistics, without culling, the national population would double every 1.6 years. Although numbers are due to significant fluctuation, government figures in 2011 estimated that there were 2.6 million feral goats living on just 28 percent of

Australia's land mass, including its coastal islands, causing an estimated $7.7 million damage to the country's agricultural industries and economy each year. As with feral populations elsewhere, goats cause significant damage to the environment, and compete with other species. But there is a silver lining. There is a strong export market for Australian goat meat, and about 90 percent of it comes from captured feral goats.

Below: Some farmers gather and sell feral goats to abattoirs, while others incorporate them into traditional farming programs. Most Australian goat meat is exported, with a high percentage shipped to ethnic markets in the US.

CHAPTER 2

Anatomy
& Biology

Anatomy of a Goat ✑

Goats are present today in large numbers on every continent (except Antarctica) where they are kept for their milk, meat, hides, and fiber, and also as working animals and pets. According to the Food and Agriculture Organization of the United Nations, there were over 875.5 million goats in the world in 2011, the last time a census was taken, and at least 570 recognized breeds. Across all breeds, however, and allowing for differences between males and females, the basic anatomy of a goat remains the same.

Right: *Shown here is a female horned goat. For more on sex-specific features, see The Mammary System (on pages 54–57) and Bucks (pages 80–83).*

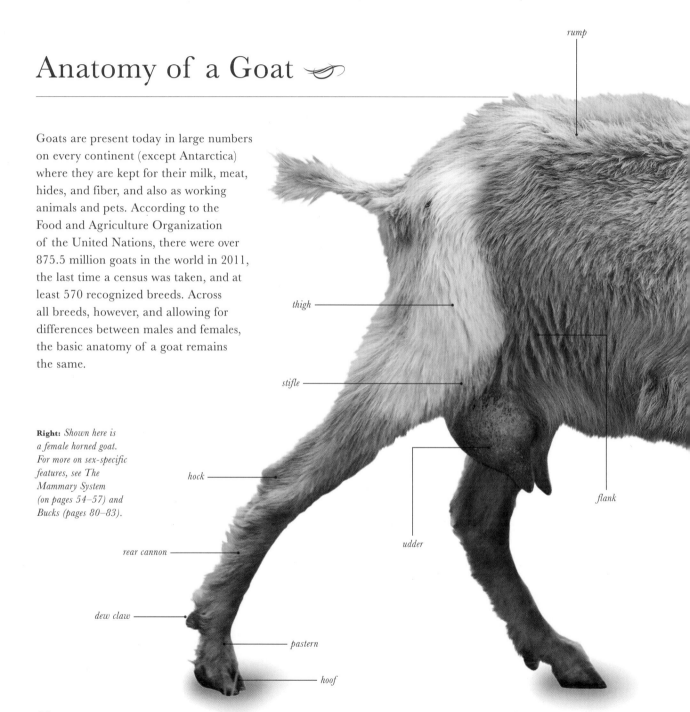

rump

thigh

stifle

hock

rear cannon

dew claw

pastern

hoof

udder

flank

withers

neck

horn

poll

muzzle

shoulder

brisket

foreleg

elbow

front cannon

GOATS AS MAMMALS

Goats are warm-blooded vertebrate animals of the class Mammalia. Mammals are characterized by the presence of mammary glands that in females produce milk for nursing their young. They have hair or fur—hair in the case of goats—and sound is carried from the eardrum by a chain of three bones. They also have two knobs at the base of the skull that fit into the topmost neck vertebra and a brain area known as neocortex—part of the cerebrum—which is associated with higher brain functions such as spatial reasoning, cognition, memory, and sensory perception.

The Life Cycle of a Goat

A goat passes through four phases throughout its life—as a kid, adolescent, adult, and senior. Exactly when it reaches the latter three stages depends a great deal on its breed, nutritional level, parasites, disease, and how it's managed, but the following provides a general overview. (See Chapter 3 for more on reproduction and early life.)

Breed strongly influences the age of puberty, with small breeds like Nigerian Dwarfs becoming sexually mature at 3 months or less, while 5 to 7 months is more the norm for larger breeds.

KIDS

Kids are born after a gestation period of roughly five months, and litters typically range from a single kid to as many as five. Healthy kids weigh between 2 and 14 lb. (1–6.4 kg), depending on breed, sex, nutrition (of the doe), and other variables.

Most feral goats and goats raised under natural conditions begin weaning their offspring at around 3 to 5 months of age, although some don't wean until their next round of kids is due to arrive. Many goat keepers separate kids from their dams at around 10 to 14 weeks.

GOAT TERMINOLOGY

Buck Sometimes colloquially called a billy; a male goat that has not been castrated generally kept for breeding purposes

Buckling An uncastrated male goat under 1 year of age

Dam A female parent

Doe Sometimes colloquially called a nanny; a female goat

Doeling A female goat under one year of age

Kid A juvenile goat, either sex

Sire A male parent

Wether A neutered male goat of any age

Right: The key stages in a goat's life. The physical structure of the senior goat shown at the right has broken down after years of bearing kids, creating a unique "saddlebags" effect.

Kid   Adolescent   Adult   Senior

ADOLESCENTS

In the US an adolescent goat between 1 and 2 years old is called a "yearling"; the British equivalent term is "goatling." However, sexual maturity is a better guide to the onset of adolescence. Most does are bred for the first time at around 1 year, thereby becoming adults.

ADULTS

Most goats are physically mature at 2 or 3 years of age. Most does produce one litter of kids a year and their milk production is at full capacity, bucks are capable of impregnating 50 does or more in a season, and working wethers can carry or pull full burdens.

SENIORS

When a goat becomes "old" depends a great deal on how it's been maintained throughout its life. A meat breed doe producing two litters every year can be old at 8, while a 10-year-old dairy doe that is well cared for may still be milking strongly and producing healthy kids once a year. Generally, a goat of nine or ten years can be considered old, while a 12- to 15-year-old, if properly cared for, will be elderly but not ancient. Other factors include the condition of a goat's teeth—especially if it has to browse its own sustenance— sex (wethers tend to live longer than does and bucks), and breed, with some breeds being noted for longevity.

Below: *Adult goats range significantly in weight and height, from tiny West African Dwarfs right up to the mighty Beetals of Asia. Many variations in coat type also occur. For more details on these and other breeds, refer to the directory at the back of this book.*

GOAT COMPARISONS

BREED	WEIGHT (male)	HEIGHT (male)	HAIR TYPE
West African Dwarf	44–55 lb. (20–25 kg)	12–20 in. (30–50 cm)	short but not sleek
Icelandic	130–65 lb. (59–75 kg)	30–32 in. (76–81 cm)	very long, slightly wavy, poofy bangs
Valais Blackneck	145–200 lb. (66–90 kg)	30–34 in. (76–86 cm)	long, straight
Golden Guernsey	minimum 150 lb. (58 kg)	maximum 28 in. (71 cm)	varies somewhat, usually long, straight
German Improved White	165–210 lb. (75–95 kg)	31–35 in. (79–89 cm)	short but not sleek
British Saanen	minimum 170 lb. (77 kg)	minimum 32 in. (81 cm)	short, can be sleek or slightly rough
Damascus	170–200 lb. (77–90 kg)	32–34 in. (81–86 cm)	long, straight, or somewhat wavy
Nubian	minimum 175 lb. (79 kg)	minimum 35 in. (89 cm)	short, sleek
Angora	180–225 lb. (82–102 kg) clipped	48 in. (122 cm)	long ringlets
Beetal	125–300 lb. (57–136 kg); 2019 world record: 683 lb. (310 kg)	minimum 36 in. (91 cm)	short, sleek

The Skeleton ✍

THE GOAT'S SKELETAL SYSTEM

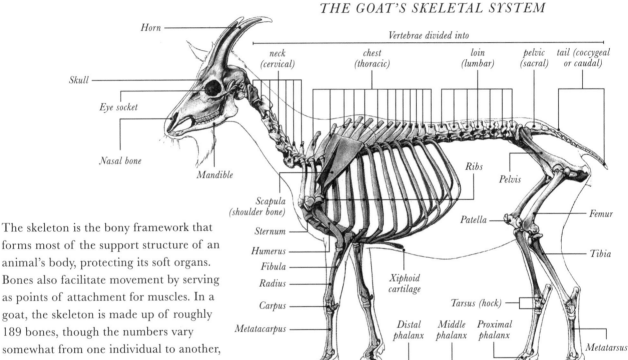

Horn

Skull

Eye socket

Nasal bone

Mandible

Scapula (shoulder bone)

Sternum

Humerus

Fibula

Radius

Carpus

Metacarpus

Phalanges

Vertebrae divided into

neck (cervical)

chest (thoracic)

loin (lumbar)

pelvic (sacral)

tail (coccygeal or caudal)

Ribs

Pelvis

Patella

Femur

Tibia

Xiphoid cartilage

Tarsus (hock)

Distal phalanx

Middle phalanx

Proximal phalanx

Metatarsus

The skeleton is the bony framework that forms most of the support structure of an animal's body, protecting its soft organs. Bones also facilitate movement by serving as points of attachment for muscles. In a goat, the skeleton is made up of roughly 189 bones, though the numbers vary somewhat from one individual to another, and between breeds.

Like the human skeleton, a goat's skeleton can be considered as having two components—the axial skeleton and the appendicular skeleton.

THE AXIAL SKELETON

The axial skeleton is made up of the skull, vertebrae, ribs, and sternum.

The skull is composed of many bones joined together by joints called sutures.

The large number of bones, with differing shapes and sizes, accounts for differences in head shapes found in individuals.

The spinal column consists of a series of vertebrae. Seven neck (cervical) vertebra allow for movement of the head and neck. These are followed by around 13 chest (thoracic) vertebrae, each of which has a pair of ribs attached to it.

Above: *Not all goats have precisely 189 bones because the exact number of ribs, lumbar vertebrae, and tail bones can vary from goat to goat.*

Six or seven loin (lumbar) vertebrae allow for flexation in the loin area of the goat. The sacrum is formed by the fusion of four pelvic (sacral) vertebrae and is usually described as a single bone. Finally, there are the tail (coccygeal or caudal) vertebrae. The number of these vary greatly from goat to goat, with some animals possessing as few as four, and some occasionally having as many as 12.

The first pair of ribs are short, straight, and thick. Near the center of the ribcage, the ribs lengthen and become more curved, and then shorter, but the curvature continues to increase. There are usually 13 pairs of ribs, most of which are joined to the breastbone (sternum) at the front of the chest and are therefore known as sternal ribs. These ribs are fixed. Sometimes there are also one or two pairs of "floating" (or asternal) ribs that are not attached to the sternum, but these are not present in all goats.

THE APPENDICULAR SKELETON

The appendicular skeleton consists of the bones of the limbs— the forelegs (thoracic limbs) and hind legs (pelvic limbs). The forelegs are comprised of a shoulder, forearm, and lower leg, while the hind legs include the pelvic girdle, thigh, and lower leg. Four digits are present in each foot. Only two, the third and fourth, are fully developed; these form the claws of the hoof. The other two are vestigial remnants containing one or two bones; they culminate in dew claws. The coffin bone, or pedal bone, is the largest bone in each claw.

GOAT SKULL

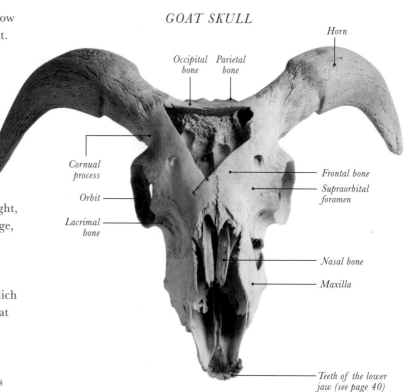

Horn

Occipital bone Parietal bone

Cornual process

Orbit

Lacrimal bone

Frontal bone

Supraorbital foramen

Nasal bone

Maxilla

Teeth of the lower jaw (see page 40)

LOWER LEG BONES

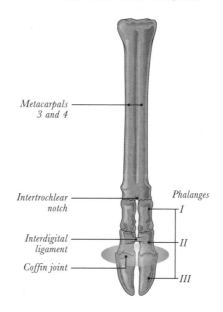

Metacarpals 3 and 4

Intertrochlear notch

Interdigital ligament

Coffin joint

Phalanges

I

II

III

Above: *A goat's skull is remarkably sturdy, designed as it is to support its heavy horns and to withstand concussion if it clashes with its peers.*

Left: *A frontal view of the bones of the lower leg and foot, showing the fused metacarpals, which diverge into the phalanges, creating the goat's cloven hoof (see page 46).*

The Ruminant Digestive System

Cellulose is a plant component that most animals cannot digest, but ruminants (goats, along with sheep, cattle, deer, bison, giraffes, and others) have a complex digestive system that is able to break it down so that the nutrients from plant matter can be absorbed. Food is taken in and then rests in the first of four stomach compartments. Later, when it's resting, the animal regurgitates this food, re-chews it, then swallows it again in a process called chewing the cud, cudding, or ruminating. Healthy goats spend up to one third of every day ruminating.

The goat's digestive tract consist of the mouth, esophagus, a four-compartmented stomach, the small intestine, cecum, and large intestine.

THE GOAT'S DIGESTIVE SYSTEM

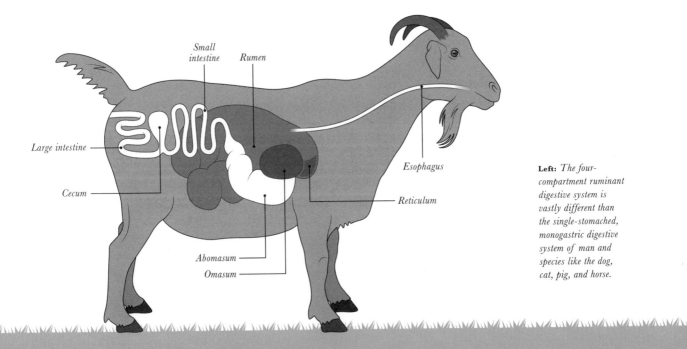

Left: *The four-compartment ruminant digestive system is vastly different than the single-stomached, monogastric digestive system of man and species like the dog, cat, pig, and horse.*

THE MOUTH

A goat takes in food with its lips, tongue, and the dental pad on the front of its hard palate (see pages 40–41). Coarser browse is chewed briefly, and grass and supple weeds are swallowed mostly whole. Plant matter is mixed with saliva to form a "bolus" before being swallowed.

ESOPHAGUS

Food next passes down the esophagus to the first stomach compartment in a series of muscular contractions called peristalsis. The ruminant esophagus is also capable of reversed peristalsis, which allows partially chewed food to be regurgitated with a gentle burp. Food is typically chewed again for about a minute, and during second and subsequent chewings,

it is mixed with more and more saliva. The additional chewing breaks the food down further, while the saliva, being alkaline, helps regulate the pH of the rumen at around a healthy 7. Goats produce up to 6 gallons (around 23 liters) of saliva in a 24-hour period.

RUMEN

The rumen, or paunch, is the first and largest compartment of the stomach, and it receives newly consumed food. It is capable of holding 3 to 6 gallons (around 11 to 22 liters) of semi-digested matter, depending on the size of the goat and type of food. It's lined with small, thin projections called papillae, which increase its absorptive surface, and it functions as a fermentation vat.

The rumen is home to billions of microorganisms that supply enzymes to break down cellulose and other feed components. This activity causes the starch and fiber in feeds to be converted to volatile fatty acids, which are then absorbed through the rumen wall and provide up to 80 percent of a goat's total energy requirements. Microorganisms also convert feed into B-vitamins, vitamin K, and essential amino acids. In the process, they produce large amounts of methane and carbon dioxide, which the goat controls by belching. The fermentation process in the rumen also produces significant heat that contributes toward keeping the goat warm.

RETICULUM

Also called the honeycomb, the reticulum is the second stomach compartment, located just below the entrance of the esophagus into the stomach. (Terminology-wise it's sometimes lumped together with the rumen and called the reticulo-rumen.) Once liquefied, feed leaves the rumen through an overflow connection known as the rumino-reticular fold, and then enters the reticulum, where fermentation continues. Material is mixed by strong contractions of the walls of both the rumen and the reticulum; this movement prevents clogging and distributes microorganisms throughout the ingested material.

OMASUM

Food next passes through a short tunnel into the omasum, or manyplies. The omasum's primary purpose is to remove excess liquid. It is divided by long folds of tissue that resemble the pages of a book and are sometimes referred to as leaves. These leaves are in turn covered with tiny, fingerlike papillae (much smaller than those in the rumen) that increase the surface area even further, for effective water absorption.

ABOMASUM

The abomasum, or true stomach, is where digestion occurs much as it does in the stomachs of humans, cats, and dogs. Like the omasum, the abomasum is lined with folds that greatly increase its surface area. Cells in the walls of the abomasum secrete digestive enzymes and hydrochloric acid, which quickly lower the pH of feed slurry from about 7 to around 2.5. Protein is partially broken down in the abomasum before slurry is shunted along to the small intestine.

SMALL INTESTINE

The small intestine is a very long tube running from the abomasum to the large intestine—20 times longer than the overall length of the goat. As semi-digested feed flows into the small intestine it's mixed with secretions from the liver and pancreas that push the pH back up to around 7, making it possible for the enzymes in the small intestine to reduce remaining proteins into amino acids, starch to glucose, and complex fats into fatty acids. Like other parts of the digestive system, the small intestine is lined with papillae that increase its absorption area.

Right: *Although some protein escapes rumen fermentation and is absorbed in the small intestines, most food that a goat consumes via grazing or browsing will be subject to the process of rumination.*

Right: *Unlike monogastric species, ruminants, especially goats, are designed to masticate and digest an insoluble fiber known as cellulose, found in leaves, twigs, and fibrous plants.*

CECUM

The cecum, also called the blind gut, is a tube between the small and large intestines, though it's often considered to be part of the large intestine. It's separated from the small intestine by the ileocecal valve, which is normally closed but opens periodically in order to allow the passage of semi-digested food. The cecum serves as a storage organ to give microorganisms additional time to digest cellulose.

LARGE INTESTINE

Muscular contractions continually push small amounts of material through the small and into the large intestine, or colon. Microorganisms continue the digestive process here, while fluid is absorbed from the slurry. Finally, fecal pellets form in the end portion of the large intestine and are secreted via the anus.

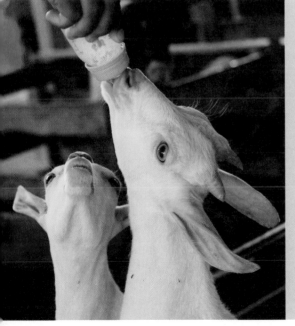

DIGESTION AND THE KID

When kids are born, their rumens are the smallest of the four stomach compartments at 30 percent of the total stomach area, while the abomasum is the largest at 70 percent (see page 89). When a kid lifts its head to suckle, milk flows through the esophagus to the esophageal groove, a muscular flap at the lower end of the esophagus that closes, creating a tube allowing milk to go directly into the kid's abomasum. This prevents milk from being fermented or soured by the ruminal microorganisms. As kids begin sampling vegetation, the rumen, reticulum, and omasum gradually develop size and function, reaching full development at about 3 months.

If the kid's head is lowered, as in drinking milk from a pan, the groove doesn't close and milk goes to the underdeveloped rumen where it's poorly digested. This is why it's important to hold the bottle of bottle-raised kids in a position replicating feeding from the dam.

Lips & Teeth ✒

Digestion begins in the mouth and goats are well equipped to gather and chew the tough, cellulose-based twigs, leaves, and plants that they feed on.

LIPS

A sheep's upper lip is divided by a deep vertical groove that helps it nibble grass close to the ground. Because goats are browsers rather than grazers, their upper lip is not deeply grooved, it is mobile. Goats are therefore able to use their nimble lips and tongue to select and rip off choice plant parts, pick bits of hay from a pile, or even separate additives they don't like from the grain in their feed. (For more differences between sheep and goats, see pages 64–65.)

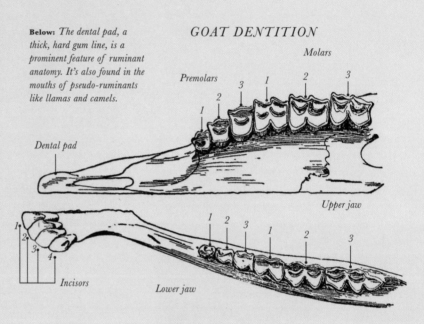

Below: *The dental pad, a thick, hard gum line, is a prominent feature of ruminant anatomy. It's also found in the mouths of pseudo-ruminants like llamas and camels.*

GOAT DENTITION

Molars

Premolars

Dental pad

Upper jaw

Incisors

Lower jaw

Left: *Goats have mobile lips that they use to gather plant material into their mouths, before biting or ripping it off and chewing.*

TEETH

Goats have three groups of teeth: four pairs of incisors (cutting teeth) on the lower jaw, three premolars (chewing teeth) on each side of the upper jaw, and three molars (also chewing teeth) on each side of the upper and lower jaws, for a total of 32 teeth. Like cattle and sheep, goats do not have incisors in the upper jaw. In their place is a smooth, hard, cartilaginous mass of tissue called a dental pad, located directly in front of the hard palate, that is used to help gather food.

It's relatively easy to determine the age of a goat by looking at its teeth. A goat's first teeth are called milk teeth. Most kids are born with their two front incisors in place; if not, they erupt within the first week of life. Second incisors then come in during the second week, the third during the third week, and then the fourth set during the fourth week. Premolars erupt between 2 and 6 weeks of age.

A first pair of permanent or adult teeth then erupts in the center of the mouth, replacing the baby teeth by around 12 months. The next two permanent teeth erupt alongside the first pair at about 1½ to 2 years of age. A third set erupts at 2½ to 3 years of age, and a fourth set erupts when the goat is between 3½ and 4 years of age. After that it's harder to tell exactly how old a goat is, though teeth elongate, spread, and break or fall out as an animal ages. A "smooth mouth" is a goat that has lost all of its permanent incisors, usually at around seven years, and a "gummer" is an old goat that has lost all of its teeth.

DENTAL MALOCCLUSION

A goat's lower teeth should meet flush with its dental pad. If they extend beyond this point the goat is said to have an underbite and is called sow-mouthed or monkey-mouthed. This is fairly common in Roman-nosed breeds. Conversely, if a goat's dental pad extends beyond its lower teeth it has an overbite and is said to be parrot-mouthed.

Goats that are badly sow-mouthed or parrot-mouthed have difficulty browsing and grazing. As a result, they tend to lose weight on pasture and so need supplementary feeding to survive.

Both conditions are hereditary and to be avoided, especially in young goats, as these conditions worsen with age. They are also both severely penalized in the show ring.

Below: Ruminants don't have canine teeth. This leaves an open space along the jaw useful for inserting fingers to pry open a goat's mouth.

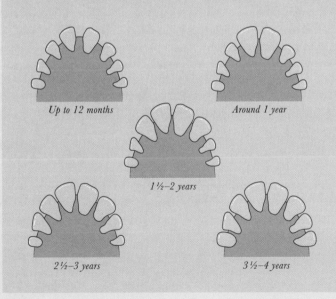

THE DEVELOPMENT OF ADULT TEETH

A young adult goat's age can be determined by its teeth, which develop in clear stages. By the time a goat is old, its gums will have receded and its teeth will appear elongated. Teeth will likely also have fallen out or become worn down.

Up to 12 months

Around 1 year

1½–2 years

2½–3 years

3½–4 years

The Five Senses ✒

The way goats perceive their surroundings through their five senses has a significant impact on their behavior—and indeed, as prey animals, their very survival. Sight, smell, hearing, taste, and touch (both separately and in various combinations) help them to learn, to sense predators, to identify foods that are safe to eat, and to establish and maintain relationships with their kind and with other species, including man.

SIGHT

Goats obtain nearly 50 percent of their information visually, making sight their dominant sense. A goat's eye is similar to a human eye, consisting of a lens, cornea, iris, retina, and optic nerve. One key difference, though, is that a goat's pupil (like that of many other hooved animals) is a large horizontal rectangle. This acts like a wide-angle lens on a camera, giving a goat panoramic vision in the range of 320 to 340 degrees—a boon for a creature that must always be on the lookout for predators. Different breeds of goat have similar but not identical fields of vision, depending on the placement of their eyes. Extensive field of vision allows goats to maintain eye contact and spatial relationships with animals both in front of

THE GOAT'S EYE

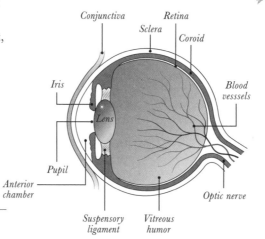

them, as well as behind them and to both sides. However, this wide field of vision reduces the goat's depth perception; it also reduces its binocular field to just 20 to 60 degrees, so they have little stereoscopic vision.

Goats are farsighted and have slight astigmatism, a common trait in prey animals. As a result, they likely perceive a well-focused image of objects in middle and long distance ranges. Goats, with their wide angle vision, are also especially sensitive to motion—highly valuable for remaining alert to predators—and a sudden quick movement will often startle them.

Left: *The components of a goat's eye are very similar to those of a human eye, but their square pupil gives them a far broader field of vision.*

Above: *Goats, sheep, deer, moose, horses, and many other prey species have rectangular pupils to increase the accuracy of depth perception in their peripheral vision.*

Right: *Prey animals like goats need panoramic vision to detect predators approaching from various directions but also to see straight ahead to flee quickly and safely.*

WHY GOATS' EYES SHINE IN THE DARK

Behind the retina in many animals, including goats, there is a colorful, shiny layer of tissue called the tapetum lucidum that gives improved vision at night. Regular vision is provided by light entering through the pupil and hitting photoreceptors in the retina at the back of the eye. The reflective properties of the tapetum lucidum reflect light back again to the retina, increasing the amount of light available to these photoreceptors. When a small amount of light is reflected off of the tapetum lucidum, say from a flashlight, a goat's eyes seem to glow in the dark.

THE GOAT'S FIELD OF VISION

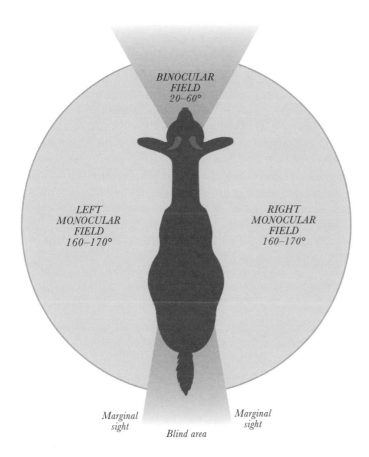

BINOCULAR
FIELD
20–60°

LEFT
MONOCULAR
FIELD
160–170°

RIGHT
MONOCULAR
FIELD
160–170°

Marginal
sight

Marginal
sight

Blind area

Goats also rotate their eyes when they lower their heads to the ground, keeping their eye slits nearly parallel to the ground at all times, no matter the position of their heads. Their eyes can rotate by as much as 50 degrees, which is ten times the ability of the human eye. This way a goat maintains a prey's-eye view of the world at all times.

Goats are dichromats with two types of cones (color-sensitive retina cells), most sensitive to yellowish-green and blue-purple light (humans are trichromats and see a full color spectrum). Goats can distinguish yellow, orange, blue, violet, and green from gray shades of equal brightness; they cannot see red and perceive it as gray or light yellow, depending on the intensity of the color. Studies have also indicated that goats can distinguish basic geometric shapes. And when shown cards depicting other goats, they recognize their herdmates and family, while when shown pictures of human faces, they choose smiling faces over glum ones (see page 94).

HEARING

Goats have excellent hearing, particularly breeds with erect ears. They shift their ears in the direction of a sound (floppy-eared breeds lift the bases of their heavy ears to do this), and are sensitive to a wide range of sounds, from the high-pitched scream of a kid in peril to low-pitched snorts or hooves stamping on the ground. A recent study concluded that goats have a broad hearing range, from 80 Hz (hertz) to 40 kHz (kilohertz), with a well-defined point of best sensitivity at 2 kHz. The human ear, by comparison, is most sensitive between 1 to 3 kHz. Within one to two hours of birth, does fully recognize their kids by sound, taste, and scent. They also recognize their own kids' vocalizations shortly after birth and remember them for life.

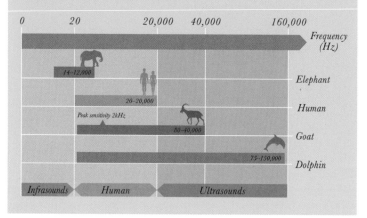

A COMPARISON OF HEARING RANGES

A goat's hearing is most sensitive at around 2 kHz. Acute hearing is key for survival, at least for feral goats and for domestic goats pastured where predators may be present.

0	20	20,000	40,000	160,000

Frequency (Hz)

14–12,000 — Elephant

20–20,000 — Human

Peak sensitivity 2kHz

80–40,000 — Goat

75–150,000 — Dolphin

Infrasounds   Human   Ultrasounds

THE FLEHMEN RESPONSE

In the nasal cavity just above the hard palate, goats have an olfactory sensory organ known as the vomeronasal (or Jacobson) organ. When a goat extends its neck, raises its head, and draws back its upper lip in a distinctive grimace, it is exposing this organ in order to draw in a new scent. This is known as a flehmen response (see left)—so named from German *flehmen*, meaning to bare the upper teeth, and the Upper Saxon *flemmen*, to look spiteful—and is held for between 5 and 60 seconds. (Flehmen is not exclusive to goats; it's performed by a wide range of mammals who also possess a vomeronasal organ.)

This behavior is used frequently by bucks, especially when they are in rut, to take in sex pheromones. But does and wethers flehmen, too, when encountering a novel smell or taste that they wish to analyze. And in does, male pheromones activate luteineizing hormone, responsible for ovulation.

SMELL

Goats are macrosmatic animals, meaning they have a highly developed sense of smell. Does use this keen sense of smell to recognize their offspring, bucks use it to determine which does are in heat, and all goats use smell to locate water from a distance and to discern musty or moldy feed. During the mating (rutting) season, bucks also exude a strong stench that both attracts does and stimulates the production of luteinizing hormone in the pituitary gland, which helps bring them into heat (see pages 80–83 for more).

TASTE

Goats have many more taste buds than humans (15,000, compared to our 9,000), and can distinguish between the same five primary tastes that we can—sweet, salty, bitter, savory, and sour. They are also surprisingly selective about what they eat. They won't, for example, drink water fouled with feces, and unless starving, they will not eat moldy or musty feed. Goats also prefer certain tastes—particularly sweets (their sensitivity to sweet is lower than that of cattle, but greater than that of sheep)—although they're more tolerant of bitters than other livestock species.

TOUCH

A goat's sense of touch is very acute. A fly alighting on a goat, for example, will elicit a strong response. Touching is also an important bonding interaction between goats, especially between siblings and between does and their kids. Goats also love being stroked and scratched by humans, which is perhaps why they can be tamed more easily than cattle or sheep.

Studies also indicate that goats have a lower pain threshold than cattle or sheep, and respond to pain more quickly. They are said to experience pain in a way that is similar to humans, but although they may feel pain (or illness) acutely, they will try to conceal this because such signs of weakness can provoke attack by predators.

Above: *Touch is important to goats, as evidenced by the physical closeness of does and their kids, family groups, and bonded friends.*

Hooves

Goats are cloven-footed mammals, meaning that they have a "divided" hoof—their hooves consist of two claws that are comparable to the third and fourth fingers of the human hand. The coffin bone, or pedal bone, is the largest of the claw bones and it gives the claw its shape. Above each hoof, at the fetlock joint, is a rudimentary, secondary hoof called a dew claw, said to have evolved as a rubbery bumper to help keep wild goats from slipping when climbing in mountainous terrain, though it has no use in today's domestic goats.

The wall, or outer covering, of each claw, is composed of keratin—horn tissue that is produced at the coronary band, the junction between the hoof and the rest of the leg. As with a human fingernail, new cells are continually produced that are then gradually pushed away and die, producing new outer growth. The weight of the goat falls on the wall of the hoof, and horn grows downward as the wall wears away during walking or climbing on hard surfaces. The soft, shiny new growth that emerges at the coronary band is called perioplic horn; it holds moisture

Above: *The underside of a goat's hoof clearly shows the division into two claws (see also page 35).*

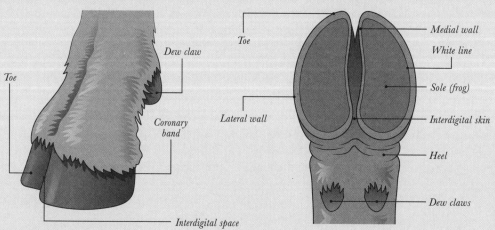

THE HOOF IN PROFILE

Toe

Dew claw

Coronary band

Interdigital space

UNDERSIDE OF THE HOOF

Toe

Lateral wall

Medial wall

White line

Sole (frog)

Interdigital skin

Heel

Dew claws

Left: *The horny sheath that forms the outer layer of the hoof is produced by horn bud cells that are located around the coronary band.*

Above: *Wild goats, feral goats, and goats that regularly traverse mountainous or rocky terrain generally keep their hooves worn down and in the pink.*

in the hoof, and its rate of production varies depending on a goat's health, nutrition, and living conditions.

The sole beneath the hoof is softer and suppler than the hoof wall, and at the rear of each claw is a soft bulb called the heel bulb. The sole is bound to the hoof wall at the white line, which forms a flexible juncture between the two.

HOOVES MADE FOR CLIMBING

The wall of each hoof is shaped like a parabola when viewed from the bottom. This parabolic shape adds strength, while the softer sole provides traction on sloped surfaces. And since each claw can operate independently of the other, a goat can use one to gain purchase on narrow surfaces, or it can splay both claws to increase contact area.

GOAT GAITS

Goats move at three standard gaits: walk, trot, and gallop. The exceptions are some myotonic goats (see pages 66–67) that are too "stiff" to trot or even gallop in a fluid manner, and specialty goats like Pakistan's renowned Nachi "dancing goats," which have a mincing dance-like gait due to their unusual conformation.

Walk Walking is the gait used most often by a goat. It is a four-beat gait with the following sequence: right hind leg, right foreleg, left hind leg, left foreleg. Goats walk at a speed of about 2 miles (3.2 km) per hour.

Trot The trot is a two-beat gait, with the legs moving in diagonal pairs. The sequence is: left foreleg and right hind leg together, then right foreleg and left hind leg together. Goats trot if they're in a hurry but not being rushed. A medium trot is about 10 miles (16 km) per hour.

Gallop Goats gallop when they are frightened or eager to arrive at a point quickly, perhaps at feeding time. The gallop is a four-beat gait. While galloping, the animal "leads" with either its left or right foreleg. The gallop sequence in the right lead is: left hind leg, right hind leg, left foreleg, right foreleg. Goats gallop at roughly 14 to 16 miles (22.5–25.7 km) per hour.

Horns

According to archaeological evidence, the earliest domestic goats had large, straight, scimitar-like horns like those of the bezoar. Twisted horns originated later in western Asia and soon predominated there, as they do today. As domestication continued, horns grew smaller as a result of selection; goats with shorter horns took up less space in pens and were less likely to injure handlers and herdmates.

A goat's horns consist of a covering of keratin and other proteins over a core of live bone that grows up from the skull. Unlike antlers, horns are never shed. Horn cores begin as tiny bony growths under the skin called ossicones, or horn buds. Each bud begins growing soon after birth, and the horn core that it becomes continues growing from its base throughout the life of the animal.

Horns grow steadily in spring, summer, and early fall, slowing considerably in later fall and winter. They don't regenerate if broken and they are never branched. Horn growth is determined by breed, sex, age, health, and the constitution of each individual. Intact males' horns are longer and thicker than females of the same breed; most wethers' horns fall midway between.

HORN TYPES

The most common horns in domestic goats today are scimitar horns, which grow straight back from the animal's skull in an arch, sometimes tipping gently outward at the ends. These can have sharp longitudinal ridges called keels running the length of the horn, as in

Below: *The central cavity of each horn is continuous with a corresponding frontal sinus in the skull. Frontal sinuses are air-filled spaces located within the vaulted frontal bone. They act as shock absorbers, protecting the brain from hard blows.*

HORN STRUCTURE

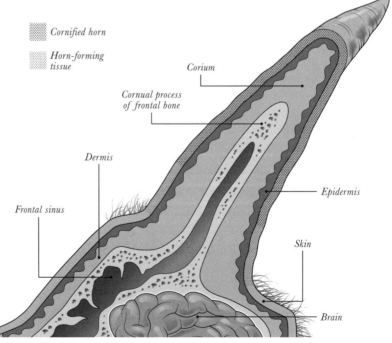

Cornified horn

Horn-forming tissue

Corium

Cornual process of frontal bone

Dermis

Epidermis

Frontal sinus

Skin

Brain

| Scimitar | Incipient corkscrew | Dorcas | Twisted |

the bucks of some dairy goat breeds, or they can be smooth like the horns of Boers, Pygmys, German meat goats, and the does of many of the European dairy goat breeds.

Dorcas horns are another type. These rise up vertically from the skull, then twist and grow at right angles to the animal's body. These are typical of Dutch Landrace, Spanish Serranas, Corsicans, and male feral goats.

A similar style is the incipient corkscrew in which the horn initially grows up and back, and then quickly twists outward to the rear, with the tips curving up and out. These are seen in Angoras, Andalusian Whites, and Canary Island goats.

Many Asian breeds have twisted horns, as does the Girgentana of Italy. The Gohilwadi and Mehsana goats of India have delicate, short horns with a mild twist, whereas the Zalawadi has larger horns with many twists along the shaft. Several Pakistani breeds have twisted horns, among them Dera Din Panah, Kamori, and Sindh Desi goats.

SCURS

Scurs are not true horns, they are horn remnants resulting from faulty disbudding (see page 90). Scurs have no core, so they break off easily, but bleed profusely. They occur most often on bucks and to a lesser degree, wethers and does. Unless kept trimmed, a scur may grow around in a circle and press against the skull or eye.

Above: *Four of the most common horn shapes found in goats. Even within these categories many variations in shape and size are seen, depending on the breed. Scimitar horns, for example, can range from the neat curves of a Boer goat to the majestic, swept-back horns seen on a bezoar ibex.*

THE ADVANTAGES OF HORNS

Horn cores and frontal sinuses are covered with a network of veins and arteries that dilate in response to stress and exercise, and constrict in cold conditions, enabling horns to function as effective thermoregulatory organs. At 86°F (30°C), a resting goat loses 3 to 4 percent of its heat production through its horns, whereas a running goat can lose as much as 12 percent of its heat production after stopping. Goats with large horns also rank higher in a herd's hierarchy than their disbudded and polled herdmates (see opposite), and removing horns from a breeding buck may make him less acceptable to does in heat. Goats tilt their horns forward to assert authority and they use them to butt, bash, or hook adversaries. They're also handy as built-in back scratchers.

Right: *Bucks of breeds like this handsome Valais Blackneck have enormous scimitar horns as impressive as those of any ibex. Does find this sexually attractive.*

Below: *While horns function primarily as thermoregulatory structures, they also come in handy for establishing a position in a herd's hierarchy, attracting mates, and scratching one's back.*

THE DOWNSIDES TO HORNS

Although horns provide convenient emergency grips when handling goats, many goat keepers prefer their animals without horns. Horned goats often get their heads stuck in fences, doors, windows, and gates—obstacles not encountered by their wild or feral kin—

POLLED GOATS

Some goats are born hornless, or polled, with just hair-covered knobs where horns would normally be. The best indicator of this at birth is a lack of swirls of hair where horn buds would normally appear; polled kids have a large central swirl on top of their heads. The polled gene is dominant, while the horned gene is recessive (see page 59), so a kid may be polled even if just one of its parents are polled. Two horned parents, in other words, will not produce polled kids. Polled to polled matings are discouraged, however, because they can result in intersex kids known as male pseudo-hermaphrodites, appearing female at birth, but developing masculine traits as they mature. They are always infertile.

and a stuck goat is at the mercy of the elements and hungry predators. Horns can also break off, and because the corneal artery supplies blood to the horn, this causes profuse bleeding. Horned goats may intentionally or accidentally injure herdmates and caretakers. And horns are used, especially by bored bucks, to destroy the sturdiest of structures.

Removing the horns of an adult goat is a horrific process that leaves gaping holes leading directly into the frontal sinuses. It requires an experienced veterinarian, general anesthesia, careful post-surgical nursing, and a long recovery time. Fortunately, it's easy to destroy a young kid's pre-emergence horn buds by way of a fairly non-invasive process called disbudding, which prevents horn growth for life (see page 90).

Above: *Though horns are beautiful they can be deadly. Many goat keepers, especially dairy goat keepers in North America and Britain, prefer disbudded or polled goats.*

Wattles ✎

Wattles—also known as waddles, bells, toggles, tassels, waggles, or lassies—are finger-like dangles of haired flesh found on the throats of goats, as well as on certain breeds of swine and sheep. Although generally found in the throat area, they are also occasionally found at other locations, such as the ears or elsewhere on the neck or head. They're usually a matched set, too, although a single wattle is not unheard of. Most are an inch or two long (2.5–5 cm) and they vary in shape from long and narrow, to short, fat blobs. Wattles occur in breeds around the world, but not all breeds of goats have them; in the US and Europe they are especially common among the Swiss dairy breeds.

WATTLES OR NO WATTLES?

Wattles serve no purpose, and goat keepers tend to love them or hate them. The latter group claims that they interfere with the fit of goats' collars, that they make a sleek throat line look awkward in the show ring,

Right: *Wattles are composed of fibrocartilage surrounded by connective tissue, covered by haired skin, and hang from the mandibular region of a goat's throat.*

that kids tend to suck on their peers' wattles causing local irritation, and that wattles are prone to getting torn off on fences and other obstructions. (Sometimes owners will secure a tight rubber band around neonatal kids' wattles so that with circulation restricted, the wattles fall off in a week or so—or, alternatively, simply snip them off where they connect to the neck; little or no bleeding occurs.) Others, however, like the look of wattles and breed specifically for them.

Above: *If wattles are to be removed, the best time is when kids are small, when little or no bleeding is likely to occur.*

SCIENCE SPEAKS

There are, perhaps, other good reasons beyond aesthetics to favor wattled goats.

Studies have indicated, for example, that does with wattles produce more milk and are more fertile. In turn, kids with wattles have greater body weights at all ages. And comparisons of bucks with and without wattles have shown that wattled bucks tend to outperform bucks without wattles in terms of body weight, height, and length, as well as chest girth and scrotal length.

Heat is a major constraint on animal productivity in the tropical belt and in arid regions of Africa. In a study of West African Dwarf goats, researchers found that goats without wattles tended to have higher temperatures and pulse rates.

Right: *Wattles are inherited via a single autosomal dominant gene, much like the gene for horns and hornlessness, so wattled kids must have at least one wattled parent.*

Below: *In the study of West African Dwarf goats mentioned in the text, researchers indicated that two-thirds of the 1,344 goats that they examined were wattled.*

WATTLE CYSTS

One downside of wattles is that wattle cysts are fairly common. These soft, round swellings form at the base of one or both wattles, or at the spots where wattles have been removed. They're ugly if large but totally harmless and often present at birth, though they don't become obvious until a goat is older. Wattle cysts refill if the fluid is aspirated using a needle, but they can be surgically removed intact. Skill is needed for this operation, however, because they lie very close to the jugular vein and carotid artery.

The Mammary System ✍

A female goat's udder consists of two separate glands, or "halves." Each half is independent of the other and delivers milk through its own teat.

Mammary glands consist of secreting tissue and connective tissue, with the number of secreting cells determining the milk-producing capacity of the udder. Bigger is not necessarily better, therefore, as a large udder might incorporate a lot of connective tissue and fat but fewer productive secreting cells than a smaller one. The secretory cells are arranged in single layers in spherical structures called alveoli. Several alveoli together form a lobule, which are in turn organized into larger units known as lobes. Milk produced by the secretory cells is then discharged into collection ducts that lead to the gland cistern, which lies directly above the teat. Milk produced between milkings is mostly stored in the alveoli but also in the gland cistern.

Each teat consists of a gland cistern, a teat cistern, and a streak canal. The streak canal is surrounded by bundles of muscle fibers that keep the teat canal closed between milkings.

Below: *This schematic diagram illustrates the components that make up the udder. Milk is produced by the cells in each alveolus. It then passes from each lobe via the ducts down to the gland cistern, from where it is released through the teat.*

CROSS-SECTION OF THE UDDER

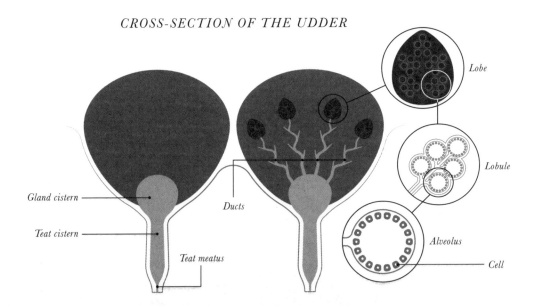

MILK PRODUCTION

Mammary development begins in the early fetal stage and continues through a doe's first lactation. Major growth then begins when the doe is in kid (becomes pregnant), especially during the later stages, which coincide with the most rapid period of fetal growth. Then, as her gestation period nears its end, the doe's udder becomes capable of producing milk.

During lactation, oxytocin, a hormone stimulated by nursing, causes the alveoli to contract and force milk out into the ducts. This is called milk let-down, and is triggered by sensory nerves along the flanks and in the teats, or by the sound of a kid crying. A doe is said to "freshen" when she kids and begins to produce milk.

Milk production peaks around the third month of lactation then gradually declines. If a goat is milked through the winter months, production falls even further but rebounds as spring approaches. Most does are rebred during their lactation period and "dried off" (taken out of production) a few months before their kids are due. Highly productive does, however, are sometimes "milked through," meaning they're kept in production as long as they're producing a sufficient amount of milk, often for several years.

Above: *Does should be able to feed their offspring. Extremely pendulous udders, deformed teats, and teats with more than one orifice make nursing difficult or impossible.*

UDDERS AND TEATS

What constitutes an ideal udder varies around the world. For example, long, narrow udders with large teats are favored in Asia, whereas capacious, shorter-teated udders held close to the body are favored in the West. But all teats should be free of defects that interfere with nursing kids or supplying milk for human consumption. Teats should be large enough to be milked by hand, and they should have a single orifice in the tip—additional, or poorly placed, orifices spray milk where the milker doesn't want it to go.

In most cases, a doe should have two teats, the exception being breeds like Boers that were originally developed to have four teats, although many Boer breeders are now selecting back to two teats to avoid the deformities that can result from multiple teats. Additional smaller teats, called supernumerary teats, are a common feature. One study found that they occurred in 30 percent of West African Dwarf does, for example, while another showed that they predominate in a population of native goats in Japan. Supernumerary teats are undesirable because if they produce milk, milking these tiny teats is difficult, and leaving them unmilked can lead to mastitis. Does with supernumerary teats are therefore usually removed from the breeding pool.

Teats should not be divided, fused, or have protrusions that would interfere with nursing or milking. Fused teats have one or two functioning milk channels and are

Right: *The supernumerary teat on this udder is nonfunctional, making it more of a blemish than a fault. Supernumerary teats are, however, hereditary and should be avoided.*

SOME POSSIBLE TEAT DEFECTS

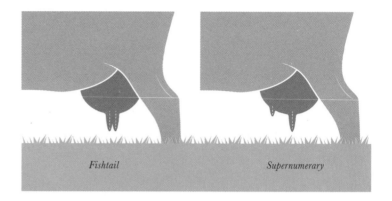

Fishtail   *Supernumerary*

Fully split   *Fused*

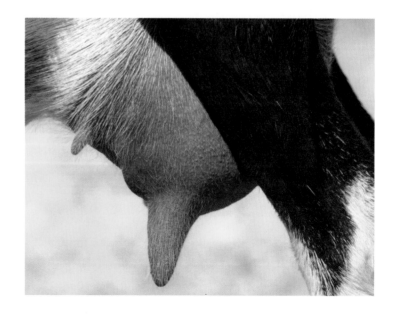

often too wide for neonatal kids to suckle. Fully split teats usually have two functional milk channels and in meat goats' smaller teats, these may resemble two normal teats placed closely together. Teats that split halfway down their length are called fishtail teats, and when both offshoots have a channel, one can spray milk into a kid's face when it suckles the other split.

PRECOCIOUS UDDERS AND WITCH'S MILK

Sometimes a doe that isn't pregnant and has never been bred develops an udder that contains milk. This is called a precocious udder. Precocious udders are not uncommon in high production lines of dairy goats. Daughters of does with precocious udders tend to develop them too, especially as they reach puberty. These udders can develop mastitis, so should be monitored daily. And occasionally, neonatal doelings are born with conical, filled teats and tiny udders that contain "witch's milk," thought to be caused by maternal hormones.

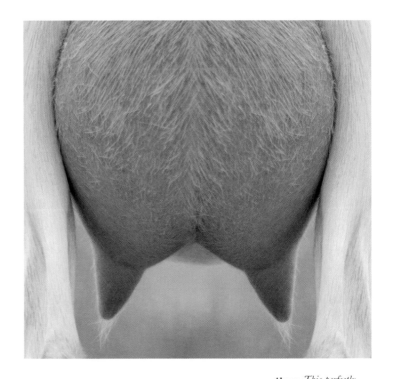

Above: *This perfectly formed udder is symmetrical, globular, and well attached rather than pendulous. It has nicely shaped teats of a size suitable for hand or machine milking.*

Left: *This Boer kid is displaying an incorrect udder—the teat on the left is a fishtail teat, and the right teat is nearly so.*

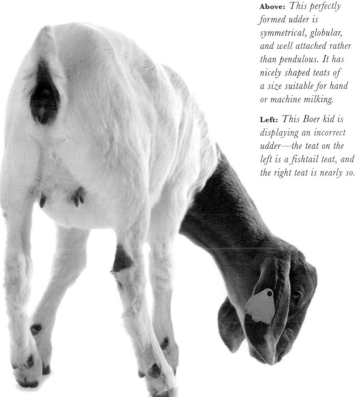

BUCKS THAT GIVE MILK

All male goats have teats but very few have udders. Although it isn't commonplace, some bucks do occasionally develop an udder directly in front of their scrotums and some of those bucks even produce milk—a condition that is known as gynecomastia.

The Inheritance of Traits

Below left: *Blue eyes, popular with Nigerian Dwarf, myotonic, and miniature dairy goat breeders, are a dominant trait, requiring at least one blue-eyed parent.*

Why goats inherit horns, particular coat colors, or any other traits is a highly complex subject. However, the subject is within anyone's grasp with an understanding of the basic principles of heredity—the way that traits are passed on from parents to their offspring.

GENETICS

Inherited traits, along with hereditary disorders, are passed down in a goat's genetic code, and the chemical basis of this genetic code is the animal's DNA, or deoxyribonucleic acid—a long molecule that is present in the nucleus of every cell of every living organism. The DNA molecule is, in turn, bound into threadlike structures known as chromosomes, with each chromosome containing hundreds or even thousands of different genes (specific sections of the DNA molecule sequence). The gene is the unit of heredity responsible for the expression of a given trait. In goats, for example, there are genes that determine eye color, coat color, size, hornlessness, or the presence of wattles.

Chromosomes transport genes when cells divide. They occur in pairs within the cell (a goat has 30 pairs), and when a sire and a dam reproduce, a kid randomly inherits one of each chromosome pair from its sire and the other from its dam—and, therefore, a copy of every gene. The trait that will be expressed, however, depends on the particular variants of these genes.

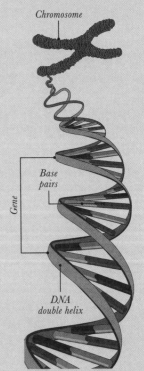

Chromosome

Base pairs

Gene

DNA double helix

DOMINANT VERSUS RECESSIVE GENES

Genes come in different variants, called alleles. They also occur at a specific location on a specific chromosome, known as a locus. So at each locus point between two paired chromosomes, a goat will have two alleles, one from each parent. If the two alleles at a particular locus are matching, the kid is said to be "homozygous" for that trait. Having two non-matching alleles for a genetic trait, on the other hand, is known as "heterozygous." The specifics of these alleles will determine which is inherited by the kid.

Alleles are dominant or recessive. A dominant allele is expressed in an animal's outward appearance (phenotype) even if only one copy, inherited from just one parent, is present. On the other hand, a recessive gene isn't physically obvious unless two copies are present, one from each parent. An animal can also be a carrier—having a recessive gene in its makeup that is not expressed but that can be passed on to its offspring.

SOME BASIC HEREDITARY TRAITS

Trait	Dominant	Recessive
hair	long	short
horns	polled	horns
behavior	nervous	tame
ears	long ears	short ears
wattles	wattles	no wattles
gender	male	female
eye color	blue	brown
myotonia (pages 66–67)	myotonia	limber-leggedness/ non-fainting
beard (sex limitation)	dominant in male goats	recessive in females

HETEROZYGOUS ALLELES

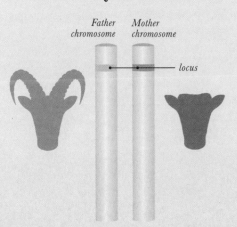

Father chromosome *Mother chromosome*

— locus

Left and far left: *If a kid inherits the horned allele from one parent and the polled (hornless) allele from the other, the kid will be polled since this is a dominant allele. And you cannot visually tell a homozygous polled goat from a heterozygous polled goat—they are both polled.*

Colors & Markings

In mammals, pigments in the hair and skin are composed of melanin, a substance produced from tyrosine, an amino acid found in the food animals eat. Two types of melanin pigments determine the color of a goat's coat: eumelanin and pheomelanin. These can be present in varying combinations. Some genes affect one or the other, and some affect both.

Eumelanin is responsible for black, dark brown, and blue-gray colors. Where a goat has eumelanin, it's solid black or brown but not a combination of the two.

Pheomelanin produces tan, cream, and red colors. These colors vary from darkest red to very pale cream.

The most important gene in goat color genetics is the agouti gene (named after the gray agouti coloration of the mouse). There are at least 14 different alleles of the agouti gene in the goat, and probably more. This gene controls the patterns of pheomelanin and eumelanin in the base coat, and the position of pheomelanic and eumelaninic areas determine the color of the goat.

Epistasis is the ability of specific allele combinations at a certain locus to mask the expression of another locus (see page 59). The masked gene is said to be hypostatic, while the gene doing the masking is known as epistatic.

This happens in goats when white masks another color and the animal appears spotted or almost solid white. White spotting is superimposed over the goat's base color—think of spotting as white areas over a dark goat and not the other way around.

A goat's final color is due to the interaction of eumelanin, pheomelanin, and white spotting. Together they form an array of colors and patterns. Some of these patterns are dominant and some recessive, and some have not been studied to date.

Below: *Gray agouti is one of the most common American Pygmy goat colors. The goat in front is also spotted and both have white caps and frosted eyes, muzzles, and ears*

BASIC COAT COLORS

White or tan · Sable · Black mask · Caramel · Bezoar

Blackbelly (badgerface) · Tan sides · San Clemente · Repartida · Peacock

SOME KEY COAT COLORS

White or tan—All tan, red, or white, sometimes with a darker shoulder or face.

Sable—Pale legs, belly, and facial stripes with a tan body.

Black mask—Tan with black on the head, brisket, and along spine. May have pale stripes on the head.

Caramel—Tan with some black on the head and lower legs, sometimes with a black belly.

Bezoar (wild color)—Tan body; a dark head with stripes, pale belly, striped legs and back. Black shoulder stripe. Males are darker than females.

Blackbelly or badgerface—Tan with a black belly; black stripe along the spine, lower legs, and face stripe. Males are darker than females.

Tan sides—Much like a blackbelly but with a wider back stripe and a nearly black head. Only the sides are tan.

San Clemente—Black front half, tan rear half; pale stripes on a dark head, and a pale belly and legs.

Repartida—Black front half, generally lacking light eye bars; a tan back half. Two-tone legs are tan in front and black at the back. Black sides on thighs. Named for a Brazilian breed of goat.

Peacock—Tan front half, black rear half; dark legs and a tan head with black stripes.

Blackbelly (badgerface)

Above: *The recognized colors and markings shown here and on pages 62–63 are based on the work of Dr. Phillip Sponenberg, Professor of Pathology/Genetics at the Virginia-Maryland College of Veterinary Medicine.*

BASIC COAT COLORS

Striped gray   Gray   Gray agouti   Swiss markings   Eye bar

Black and tan   Fishy   Lateral stripes   Mahogany   Red cheek

Striped gray—Mixture of black and white hairs over the body with pale ears, eye bars, muzzle, and legs.

Gray—Uniform mix of black and white hairs with a darker head and legs.

Gray agouti—Uniform mix of black and white hairs with distinctly darker legs.

Swiss markings—Dark body, dark belly; pale legs, ears, facial stripes.

Eye bar—Black with tan belly; rear legs black on front and tan in back; front legs tan on back with tan around leg above knee. Wide, light stripes on face.

Black and tan—Black with tan belly; rear legs black in front and tan on back; front legs black on front with tan in back and tan all around legs above knees; light inside ears; thin, light face stripes.

Fishy—Black with front half of belly tan and rear half black. Black udder or scrotum. Medium-wide bars on face. Continuous black stripes down fronts of legs.

Lateral stripes—Like "black and tan" but darker belly and reversed leg stripes (black in back).

Mahogany—Dark mix of black and tan hairs; dark legs, head, minor striping.

Red cheek—Black with tan patches on cheeks, backs of thighs, tops of ears.

No pattern—Black (not illustrated).

Above: *Goats come in a wide array of colors and markings. Some, like Swiss markings, are seen in numerous breeds, while the striped gray, for example, is far less common.*

Swiss markings

Belted   Spotted   Barbari   Flowery

Roan   Goulet   Algarve   Nigerian   Frosted

WHITE MARKINGS

All of the colors listed on the previous
pages can be modified with any of the
following forms of white spotting:

Belted—Anything from a ring around the
goat's barrel to a nearly white goat with a
colored head and tail. Likely dominant.

Spotted—Random white spots, usually
with white on the legs and body and white
spots on a darker body. Mode of
transmission is uncertain.

Barbari—White areas on face and sides
in which small dark spots remain. Named
for the often similarly colored Pakistani
Barbari goat. Probably recessive.

Flowery—Small white flecks throughout
the coat, especially the sides and belly.
Probably dominant.

Roan—A mixture of individual white
hairs into the base coat color, usually on
the body but not on the head and legs.
Probably dominant.

Goulet—White ears, color over eyes,
white lower face, and ragged-edged
white and colored spots over the body.
Probably dominant.

Algarve—Ragged-edge white and
colored spots on the body, with dark
ears and dark eye patches.
Likely dominant.

Nigerian—Dark legs,
white with roan spots
on dark areas.
Homozygotes may be
very pale and lack body
spots. Probably dominant.

Frosted—Roaning on the ears
and muzzle. Present in many
breeds but especially common
with Nubians and American
Pygmy goats. Dominant.

Above: *Roans, spotteds,
belted, and frosted
ears and muzzles
are everywhere, while
markings like goulet,
algarve, and barbari
are less commonly seen.*

Barbari

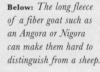
Goat

Sheep versus Goats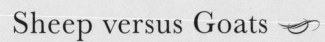

Sheep

Apart from wild goat species, sheep (*Ovis aries*) are a domestic goat's closest relatives. They have much in common, though there are striking differences too, especially in their physiology and behavior. One thing most people don't know is that goats and sheep can interbreed but because they have a mismatched number of chromosomes, ewes bred by bucks cannot conceive, and does bred by rams lose their pregnancies after five to ten weeks.

IS IT A SHEEP OR A GOAT?

It's easy to tell typical wool sheep and goats apart, but a bit harder if you factor in long-fleeced fiber goats like Angoras, Russian Dons, or long-haired breeds like Golden Guernseys, Appenzells, and Corsicans. And not all sheep are typical wool sheep; hair sheep breeds (which grow coarse hair instead of wool) look very goat-like indeed.

When in doubt, look at the tail. Goats have short tails that they carry away from their bodies, often flipped up and over their backs, unless they're frightened, worried, or sick. Sheep have longer tails that are sometimes docked (shortened) but that always hang down.

Check for a beard. Many, though not all, goats of both sexes have them but sheep do not.

Horn shape is important, too. Not all sheep have horns, but those that do mostly have horns that grow up, back, and down into spirals alongside their faces. Goats' horns mostly grow up and back, away from their faces, though the scimitar horns of older animals may grow in a single loose curve around and back beside their heads.

Below: *The long fleece of a fiber goat such as an Angora or Nigora can make them hard to distinguish from a sheep.*

Angora

Nigora

If still in doubt, look at the upper lip. In sheep it's divided by a deep groove that enables them, as grazers, to feed close to the ground. Goats, being browsers, don't need it. Note, too, the tear-shaped infra-orbital gland at the lower edge of a sheep's eyes. Sheep nuzzle this gland on one another to establish social rank.

And if it's late fall or early winter, when rams and bucks are in rut, scent can be a powerful giveaway. Rams of a few breeds will smell slightly musky, but bucks of every goat breed will reek of musk and urine.

BEHAVIORAL DIFFERENCES

Goats prefer to eat at chest level or higher and they're willing and able to stand on their hind legs to browse; sheep graze close to the ground. (Goats will graze if browse is not available, however, and hardier sheep breeds browse when grass is in short supply.)

Goats are a laying-out species—most does leave their kids in a safe spot, coming back periodically to feed them—whereas sheep are a following species. Lambs shadow their dams and stay with them whenever the flock moves.

Goats are more active, vocal, curious, and independent than sheep. Their natural wanderlust and climbing ability makes them hard to keep inside of fences, and they are more likely to separate from a herd. Sheep have a strong flocking instinct, so they are easily held in place. Yet many owners find goats—and particularly the more placid dairy breeds—easier to handle when worming, vaccinating, or hoof trimming, because frightened sheep are strongly wired to run from perceived danger.

SHARED CHARACTERISTICS

Life span usually 10–15 years

Temperature 101.5–104.5°F (38.6–40.3°C)

Pulse 70–80 beats per minute

Respiration 12–15 breaths per minute

Ruminal movements 1–3 per minute

Goat's upper lip

Sheep's upper lip

Below: *Sheep and goats are physically and temperamentally similar enough that they usually coexist nicely.*

Right: *The deep groove in a sheep's upper lip, for grazing right down to the ground. Goats' lips are shallowly grooved.*

Myotonic Goats

Myotonic, or fainting, goats are an American exclusive. Today's myotonic goats trace their lineage to three does and a buck brought from Nova Scotia to Tennessee in the early 1880s by an itinerant farm laborer. These goats became the talk of the hills because they stiffened and often fell over when startled. They piqued the interest of a Dr. H. H. Mayberry so much that he offered to buy them. Mayberry raised kids from his fainting goats and sold them to farmers throughout the area. Gradually they spread throughout the Southern states, then during the 1930s and 1940s, they made their way to Texas, where they evolved as bigger, meatier goats. Over time, though, their numbers dwindled until, in 1988, they were added to the American Livestock Breed Conservancy's Conservation Priority List and officially declared an endangered breed. Since then the organization, now called the Livestock Conservancy, has moved them to "Recovering" status.

FAINTING GOATS DON'T ACTUALLY FAINT

Myotonic goats don't actually faint; they're affected by a genetic disorder called myotonia congenita that, when the goats are startled or scared, causes skeletal muscles, especially in their massive hindquarters, to contract, hold, and then slowly release. Episodes are painless and the goats remain awake (they often continue chewing food they have in their mouths). They remain down for five to 20 seconds until the stiffness passes.

Right: Myotonic goats come in many shapes, sizes, colors and coat styles, ranging from 35-lb. (16-kg), smooth-coated Mini-Myotonics to 200-lb. (90-kg), shaggy Tennessee meat goats.

Right: When goats "faint," they stiffen and topple to the ground, usually landing on their sides.

IT ISN'T JUST A GOAT THING

Myotonia congenita isn't just found in goats—it can affect dogs, cats, tumbler pigeons, horses, mice, water buffalo, and humans too. The disorder affects an estimated 1 in 100,000 people worldwide, occurring either as the most common form—Becker disease—or the rarer Thomsen disease. Becker disease causes more pronounced muscle stiffness than Thomsen disease, particularly in males. Many people with Becker disease also experience temporary attacks of muscle weakness, particularly in the arms and hands. Although myotonia can affect any of the body's skeletal muscles, it occurs most often in the legs. While myotonic muscle stiffness can interfere with movement, the condition isn't painful.

MYOTONIC GOATS TODAY

Myotonic goats are easy to keep. They are more parasite-resistant than most other breeds, and are also kept inside fences more easily because stepping up on a fence or obstacle causes them to stiffen. There are several sub-breeds of myotonic goat. The Tennessee meat goat is a large, highly muscular animal developed for the meat market, while Mini-Myotonics are small versions kept as pets, as are Miniature Silky Fainting Goats, long-haired miniature goats developed to resemble Silky Terriers. (See pages 185 and 201 for physiological details.)

Dwarfism in Goats

Goats standing less than 20 in. (50 cm) at their withers are classified as dwarf goats. Although a number of dwarf breeds occur around the globe, today's increasingly miniature goats descend primarily from two African breeds: the West African Dwarf and Southern Sudan Dwarf.

AFRICAN ANCESTORS

In 1847, the German biologist Carl Bergmann espoused a theory that has come to be known as Bergmann's rule: that larger species are found in colder countries and smaller species in warmer ones. The theory is that the ratio of surface area to volume is higher in smaller animals, thereby enabling them to lose more heat through their skin and stay cool in warmer climates. In line with this rule, several types of dwarf goats and sheep have evolved in the hot zones of Central Africa.

First and foremost of these is the tiny West African Dwarf goat (see page 167 for a full profile), which is the product of achondroplasia—a form of genetic dwarfism that results in retarded growth of the long bones. Thus the West African Dwarf (WAD) has a disportionately plump, wide body with stout, short legs

Above: *Some breeds are based on Pygmy goat genetics crossed with those of other breeds, like the Pygora, with its mixed American Pygmy and Angora ancestry.*

and a short, wide head. It stands around 12 to 20 in. (30–50cm) tall. WAD goats are resistant to disease spread by the tsetse fly, and they're less affected by potentially fatal barber pole worms than other breeds (see page 107), so they are the only goats commonly encountered in 18 countries of West and Central Africa. The WAD is the primary ancestor of the American Pygmy goat and its own descendant breeds—Pygora fiber goats and Kinder meat and milk goats.

Above: *Miniature breeds developed in North America and Europe are descended from hardy African goats like this West African Dwarf.*

DWARF GOATS AROUND THE WORLD

The British Pygmy Goat Society, the Pygmy Goat Club of Ireland, and the Pygmy Goat Society of Ireland include both types of African-derived dwarf goats in their herd books, lumped together under the term "Pygmy goat." Australian Miniature Goats also include animals of both types. In North America, however, Pygmy goats and Nigerian Dwarfs are registered and promoted by separate organizations.

Miniature goats are popular because they're more easily handled by children, the elderly, and vulnerable adults. They require less space and less feed than standard-size goats, too, and in many cases, with their size factored in, they're more productive. For example, it's possible to house and feed four Nigerian Dwarf does producing up to a gallon of milk per day in the space and on the feed required to keep a single full-size dairy doe.

The second primary player is the Southern Sudan goat, a product of pituitary hypoplasia, a hereditary condition that results in a small but normally proportioned goat. It typically stands 16 to 20 in. (40–50 cm) tall. There are several types, the smallest being the tiny Yei or Dinka goat, and all are adapted to life in the arid and semi-arid parts of East Africa. These are the forebears of Nigerian Dwarf goats, which, in turn, led to descendants including Mini-Saanens, Mini-Sables, Mini LaManchas (Mini-Manchas), Mini-Nubians, Mini-Alpines, Mini-Oberhaslis, Mini-Toggenburgs, and Mini-Golden Guernseys, as well as Miniature Silky Fainting Goats and Nigora fiber goats (see Chapter 6 for breed profiles).

Left: *American breeders developed Mini-Manchas by crossing Nigerian Dwarf bucks with full-size LaMancha does.*

Right: *A Miniature Silky Fainting Goat, one of several breeds that descended from the Nigerian Dwarf.*

CHAPTER 3

Society & Behavior

The Goat Lifestyle

Goats are highly gregarious animals and they prefer to stay close together—except for does that are about to kid, individuals rarely stray from their herd. Within the herd, too, smaller subgroups form, thereby increasing the cohesion of the herd and decreasing instances of infighting.

LIVING TOGETHER

Social groups consist of both family and friendship groups, and the activities of individuals within a group are often highly synchronized. Family groups may include a doe, her kids, her grown wethered sons, and her daughters, along

Below: Goats are ultra-social creatures, so they should be kept together with other goats or with comparable prey species such as sheep, camelids, horses, or donkeys.

with her daughters' kids; littermates stay close for life. Friendships form when goats (particularly juveniles) are raised together or kept together separately from the main herd for a time. Members of friendship groups can be as devoted to one another as those of family groups. Signs of affection between members include resting in close proximity, reciprocal grooming, sniffing or rubbing one another, and showing distress if out of sight of a family member or friend.

A goat totally isolated from its herd becomes deeply distressed. It initially rushes around and bleats high-pitched distress calls but eventually becomes less active, quieter, and depressed. Studies have shown that goats don't become habituated to repeated isolation sessions. Stress leads to a drastic reduction in immune system function, which is why domestic goats that are removed from the herd—perhaps due to illness, injury, or prior to kidding—must be kept within sight and sound of other goats, particularly members of their own herd.

Some breeds are more innately social than others and have a strong flocking instinct—for instance, the Nachi dancing goats of Pakistan. These are shown in "Nachi walk competitions" in which a showman leads the head goat by its ear and a large group of unfettered goats follows behind.

Above: *Once goats establish their niche in their herd's hierarchy, little serious infighting occurs. Dominant goats will sometimes bully lower-ranking members but this is simply part of a goat's life.*

DEALING WITH CLIMATE

Goats in temperate climates grow a thick, wintertime undercoat of cashmere that keeps them warm in dry cold, especially if they can get out of the wind. However, wind, low temperatures, and rain can be a deadly combination, leading to pneumonia and death, so goats will seek shelter from wind and rain (although they are happy to go about their lives in dry cold, often venturing out to graze on the coldest days).

In hot, humid conditions, goats pant to circulate air through their bodies to help cool down. They seek shade, especially during the heat of the day, and often change their feeding patterns to browse at night (see below). Goats living in hot, dry, and treeless areas congregate and huddle together during the midday heat—this allows them to reduce their exposure to direct and reflected heat from the sun.

FEEDING BEHAVIOR

Many factors will influence the feeding behavior of goats, including grazing management practices, the season and the type of vegetation available, breed, group size, and more.

Goats are browsers, but they will also graze when necessary. They normally graze and browse for six to eight hours

Above: *Unlike their cousins, the sheep, goats hate getting wet. In mixed herds, sheep continue grazing in a downpour while goats will huddle miserably under shelter.*

a day, breaking feeding times into about five or six separate periods, with time out for ruminating (see pages 36–39) and resting. During summer, the browse consumed during the first few hours after daybreak usually comprises the largest single meal of the day. In this early morning session, goats tend to eat a lot and are less selective. A second large feeding period occurs in late afternoon until about sunset, with minor periods during other parts of the day. And during hot weather, goats graze more at night. In winter, most feeding occurs from mid-morning to mid-afternoon when temperatures are warmest.

Goats steadfastly reject plants, hay, and grain contaminated with the scent of their own species' urine or manure. This is a good thing as it helps prevent them from ingesting parasite eggs.

Above: *Goats on pasture alternate between grazing and resting throughout the day. While resting, they relax in groups with family and friends while regurgitating and re-chewing their cud.*

Right: *Goats have a lower pain threshold than other prey species and they're quicker to vocalize it, too. Pain and distress calls are loud and unique. Heard, they demand investigation.*

THE IMPULSE TO RUN

Goats that are terrified or in deep pain react by screaming loudly and attempting to escape. Sometimes they go into a catatonic state. This response was first recorded by Ivan Petrovich Pavlov, the Russian scientist famous for his work with conditioned responses in dogs. He believed that the response was related to inhibition of the goat's self-protective impulse to run.

Hierarchy

Every group of goats, whether it
consists of two or two hundred members,
establishes a strict, lineal social hierarchy.
And once that order is established, it
generally stays that way unless a new goat
enters the herd, whereupon it must battle
its way to its spot in the new hierarchy.

THE HERD KING

Most of the year any bucks in the group, including the alpha male, or herd king, defer to the herd queen. However, as rut begins, the herd king assumes leadership of the herd. The herd king breeds all of the does; underling bucks are simply out of luck, although bold bucks will constantly challenge the king. A herd queen, however, generally outlasts many youthful kings. The aged buck shown here was the herd king of the Bilberry goats, a breed of feral goat believed to have lived on Bilberry Rock in Waterford City, Ireland, for hundreds of years.

THE GOAT PECKING ORDER

The goat highest in the hierarchy is usually a wise, older doe known as the queen. The herd queen eats what she wants at the feed trough, claims the best sleeping place, and enters the milking parlor first. In the wild, a herd moves because the queen leads it. When she stops, they stop. When she eats, they eat—and when possible, they choose whatever she is eating. The queen can also threaten, butt, bite, or horn any other goat she pleases. She's the boss.

The next highest goat can do those things to any goat except the top-ranked individual. Third in line can harass any but the top two, and so on down the line until we reach the bottom—a shy, picked-upon individual harassed by all, with no one it can pick on.

Dominant does tend to produce dominant kids. Nursing kids assume their mother's position in the hierarchy until they're weaned and fight to claim their own place in the order, although mothers always maintain a dominant position over their kids. Kids begin seeking their own place in their herd's social hierarchy at around 6 months of age.

Goats in the hierarchy occasionally choose to move up in position by challenging a superior but this is fairly rare. And the queen is rarely if ever unseated until she's old and feeble, when another high-ranking goat—often one of her daughters—may challenge her and take her place.

Left: *Any goat in any hierarchical position in a herd can challenge the leaders of the herd to quickly advance through the ranks, though in reality they rarely do.*

FIGHTING FOR DOMINANCE

Apart from minor squabbles between family members and friends (see below), peace generally reigns in the herd because each goat knows its place and reinforces that place by threatening but rarely engaging with those that are lower in rank. It may warn them by issuing a horn threat: stretching its neck, lowering its head with its horns or poll jutting forward, and glaring. If the other goat backs down, the higher-ranking goat may chase it away. If that doesn't work, the dominant goat may press its horns or forehead against the other goat, rear without actually butting it, bite its ear, or ram its opponent's rear end or side (see box opposite).

Apart from fighting, dominance within a herd is influenced by aggressiveness, breed, sex, age, body mass, and horn size, with the latter two factors being the most important. One study found that commercially housed dairy goats engaged in more horn threats but less physical contact than their hornless herdmates, ostensibly because underlings choose to give way to horned superiors rather than feel their wrath.

Below: Glaring, head pushing, horn threats, and body shoving are mild ways in which goats assert dominance. Even hornless goats perform horn threats by advancing their polls at an adversary.

Above: *Kids assume their dam's position in the hierarchy until they're ready to fight for one of their own. In the meantime they hone their skills by play-fighting among themselves.*

Fights will often ensue when a new goat joins the herd and challenges the existing order, but fortunately, new positions in the hierarchy are resolved within days and serious injuries rarely occur. Other goats generally remain mere observers whenever a battle for dominance occurs, although they may occasionally enter the fray, especially if one of the challengers is a newcomer. But while distressing to their human caretakers, engaging in agonistic (fighting) behaviors to establish rank is a vital part of being a goat.

BUTTING

Butting tends to be practiced mostly by high-ranking members of the hierarchy. Butting can be brutal, particularly the rear clash, where the aggressor butts her target, hard, from behind. If the other goat doesn't give way, serious fighting may ensue. Antagonistic goats will stand roughly 3 to 6 ft. (1–2 m) apart. They then rear up with their bodies at right angles to one another, before pivoting, lunging forward, and bashing their foreheads together with a loud crack.

Bucks ‿❧

Adult bucks can weigh from 50 to 500 lb. (18–227 kg) and more, depending on the breed. They are hormonally driven, so even normally well-mannered bucks require careful handling, and the male goat's courtship behavior is an eye-opener to anyone new to goats. Bucks, in short, do not make good pets, for reasons we'll soon see.

BUCK AGRESSION

In nature, bucks fight for everything, from feed and shelter to breeding rights, so they're wired to be more antagonistic than does or their castrated peers. Most bucks occasionally challenge their handlers and some do it much of the time. In addition, some breeds and bloodlines are more aggressive than others, but in every case it's important to carefully raise bucklings destined for breeding status (see pages 110–11).

THAT SMELL

Although male goats are thought to smell bad, it is specifically bucks that are the culprits—whenever it is mating (rutting) season. Part of the overwhelming stench of bucks in rut is the result of spraying urine on their beards, faces, and front legs until they're impregnated with its scent (self-enuriation or scent urination).

A buck's penis has an S-shaped curve in it called the sigmoid flexure that allows it to lengthen up to 12 in. (30 cm) during mating. The urethra extends as a wormlike structure (the urethral process, or pizzle) to around an inch (2–3 cm) beyond the end of the penis.

Below: Most bucks are friendly and docile when out of rut, but being strongly hormone driven, their attitudes can change radically as breeding time approaches.

THE MALE URINARY TRACT

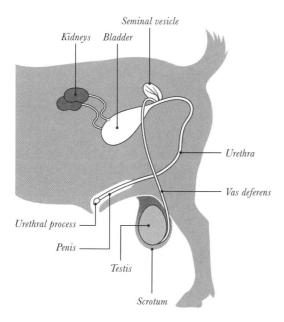

Kidneys

Bladder

Seminal vesicle

Urethra

Vas deferens

Urethral process

Penis

Testis

Scrotum

Above: *A buck's slender urethra ends in a stringlike structure called the urethral process. Calcium deposits can sometimes lodge in these organs, creating potentially fatal blockages (see page 104).*

Above right: *A buck's urethral process twirls in a corkscrew motion, ostensibly to distribute sperm in a doe's vagina but also to distribute urine throughout his environment.*

When breeding or urinating, the urethral process twirls in a corkscrew motion that distributes sperm within a doe's vagina. Designed as he is, a buck can shoot urine a long way.

To do so, he turns his head and shoulders down and toward his hindquarters, and shoots urine onto his face and front legs. Urine can be delivered as a spray or a precise stream and released steadily or in spurts. With his body curved, the buck is able to place his nose and mouth in the stream and he may lap urine with his tongue. He might end the display by grooming his penis with his tongue. Then he throws back his head and flehmens (see page 44).

By the end of rut, bucks are stinky and sticky, with a thick layer of dried urine on their faces and legs. This can cause urine scald leading to hair loss, irritation, and

open sores, particularly on their lower faces and the backs of their front legs.

The other contributing element to the smell of a buck in rut is an acrid odor exuded by scent glands located near the horns. (When a buck tries to scrub you with his forehead, he's marking you with his scent. Does and wethers also have scent glands but they don't produce the odiferous musk associated with bucks, nor do they self-enuriate.) Researchers recently discovered that one of the scent pheromones given off by bucks in this way causes a hormonal chain reaction in a doe's brain.

Buck stench intensifies as a buck ages. Most yearling bucks in their first rut have little odor. Since odor attracts does and most show a preference for bigger, older bucks, yearlings may be rebuffed when they try to breed.

Left: *In the wild, a single buck leads the herd and impregnates female members. Young bucks form nearby bachelor bands until bold enough to challenge a head buck's position.*

Below: *Bucks gobble, stomp, and whoop while insistently pursuing does until they submit. To lessen the drama, many goat handlers restrain the does they want to be bred.*

MALE MATURITY

Depending on breed, genetics, and how they're managed, some young bucks—or bucklings—are capable of inseminating does, including their mothers and sisters, while still at their mothers' sides. Live sperm has been observed as early as 110 days after birth in the epididymis of the dairy goat (the tube that connects the vas deferens to the testes), and this is the rule rather than the exception in breeds like the Nigerian Dwarf (see page 194).

THE COURTSHIP RITUAL

When a buck homes in on a likely doe, he'll walk or run beside her as she leads him around the pasture or pen, with his head alongside hers, kicking one front leg stiffly forward, flapping his tongue, and blubbering. Blubbering can be any sort of whoop, groan, or growl, but often sounds as though the buck is saying *wup-wup-wup*. If she's amenable to being bred, the doe will stop and plant her feet. The buck then sniffs her sides and her perianal area. If she urinates, and she probably will, he'll lap at her urine. If she still stands, he will mount her. He'll thrust a number of times, and then throw back his head as he ejaculates, before sliding off, resting briefly, and then trying again.

Does

Adult does can weigh from 40 lb. (18 kg) or so for a small Nigerian Dwarf or West African Dwarf, up to 300 lb. (136 kg) or more for some of the giant Asian breeds. While bucks are necessary for breeding and wethers make good pets and working goats, most of the goats kept around the world (for milking) are does, and they outnumber males many times over.

DOES IN HEAT

From early fall through mid-winter, seasonal breeders (like does from breeds that originated in Europe and North America) cycle, or come in heat (estrus),

every 18 to 23 days; aseasonal breeders (the desert or equatorial breeds that originated in hot climates) cycle all year round.

Does are stimulated by the appearance and scent of a buck (see page 80). Given their choices, most does choose an old buck with impressive horns over a younger rival. They rub their necks and bodies against him to court his favor. A fully receptive doe stands with her head slightly lowered, her legs braced, and her

Below left: *Dairy does are selected for good mammary systems, milk production, and, in large-scale dairies, the ability to coexist with peers in confinement.*

Below: *Does in heat sometimes behave like bucks, gobbling, pawing, and mounting herdmates, especially if herdmates are also in heat and no buck is present.*

SIGNS OF HEAT

- Interest in nearby bucks

- Loud and strident vocalizing

- Increased urination

- Mounting other does and wethers and being mounted by them

- Lowered milk production

- Tail wagging (also called flagging)

- Decreased appetite

- Fence walking (pacing along the side of a fence)

- Sometimes a slight quantity of a clear vaginal discharge

- Personality changes—for good or bad

tail to one side; she may urinate when he sniffs and nuzzles her under her tail.

Estrus generally lasts for two or three days. The time between the start of one estrus and the start of the next is called the estrous cycle. Ovulation occurs 12 to 36 hours after the onset of standing heat. Standing heat indicates that a doe is receptive to being bred. As does come into estrus they may run away, or at least avoid a buck if he tries to mount them, but once they're in standing heat they (usually) don't.

If impregnated, a doe enters anestrous and most does stop coming into heat, although a small percentage of does will allow themselves to be bred throughout their pregnancies.

LADIES' CHOICE

Does can be surprisingly picky about the bucks that breed them, a trait that can dismay goat owners who choose a specific mate for each doe. Does prefer bigger, older males with impressive horns and if they have a favorite they may strongly resist any other buck, especially young ones and polled or disbudded bucks, even when strongly in heat. And sometimes they reject even seemingly perfect big, old, nicely horned bucks—just because. Then the only solution is to hand breed, with a handler restraining the doe so the buck can do his job.

Above: *Some does in heat can be obnoxious indeed, crying loudly and stridently, racing around wildly, and behaving totally out of character for the duration.*

Early Life ~

Nursing kids assume their mother's position in her herd's hierarchy (see pages 76–79), so low-ranking does take pains to keep their young away from herd leaders, who are not at all patient with another doe's offspring and will rebuff them by nipping them or head-butting them away. Does also generally have favorites in litters of two or more, with preference usually given to male kids.

BONDING

As soon as a kid is born, a doe will get to her feet and begin cleaning it. It's during this process that the doe becomes bonded to her kids, so they should be left alone until that bond occurs.

Allowing for breed differences, bonding occurs within one to four hours (a significantly shorter period than for sheep and cattle), and they recognize her among others in the herd between 24 and 48 hours after birth. Kids that are removed from the birthing pen or somehow get pushed to the side or otherwise lost before the doe cleans them are sometimes rejected, although most does will still accept their own kids if returned within two hours of kidding. A kid who approaches a nursing doe that isn't its dam will be strongly rebuffed.

SUCKLING

Neonatal kids are precocial, which means that they will stand and seek their dam's udder shortly after birth. A normal kid stands within ten to 30 minutes and it will suckle within an hour. If there are more than two kids, the stronger individuals will secure teats and the weaker ones won't, so without intervention, particularly weak kids may die.

Kids kneel to nurse, and they bunt their mother's udder with their noses prior to and during nursing, an act that facilitates milk let-down. A rapidly

Above: *Most does are instinctively good mothers, rising as soon as their kids are born and cleaning them, an act that aids in permanent bonding.*

wagging tail means a kid is suckling milk. After feeding, contented kids take naps. A kid that constantly cries or continually suckles or probes at its dam's udder isn't finding enough to eat. At first, does are attentive mothers, allowing kids to nurse whenever they please. As kids grow older, their mother will let them nurse for a short time and then walk away when she feels they've had enough. This is an early step toward weaning, which doesn't occur until kids are at least 3 months old, but often much older—around 5 or 6 months. Some does don't wean until their next litter of kids is due.

Some does will also accept and raise another doe's kids, especially if they're introduced to her soon after birth. Mother and daughter and sister combos sometimes share their kids, too, with each party nursing whichever kids approach her.

Above: Kids rise soon after birth and seek their dam's udder. Problems arise when the number of kids exceeds available teats. Some goat keepers routinely bottle raise excess kids.

COLOSTRUM

Colostrum ("first milk") is a thick, yellowish milk produced by does for about 48 hours after kidding. It's packed with important nutrients, but more important, it contains antibodies that, because a kid's immune system isn't fully developed until it's around 7 weeks old, it needs in order to survive. Colostrum needs to be consumed within the first 24 hours of life, but kids should ideally ingest it within an hour or two after birth—antibody levels in a doe's colostrum decrease dramatically after around 12 hours, and the lining of a kid's intestine can only absorb those antibodies for roughly 12 to 24 hours after birth. Colostrum is so important that kids that don't ingest it rarely survive.

Left: Dams teach their kids to drink water as well as to graze and browse, pointing out which plants are suitable for eating and which are best avoided.

Below: Even domestic does sometimes hide their newborn kids. Kids are wired to remain where they're placed and to stay quiet until their dam returns.

LYING OUT

Goats are a "lying out" species; does place their newborn kids in what they deem safe spots, then go off to graze, returning roughly four to six times a day to feed their hidden young. Some, although not all, domestic does do this too. Sometimes a doe may misplace her kids, which will result in a lot of screaming and rushing about until they are located again.

Kids are wired to remain where they're placed by their dam; they can fit into very small spaces, and they will seek hidey holes for themselves. This behavior continues for a day or two, or up to a week or more, at which point kids begin following their dam out to feed, nibbling at pasture and browse. She guides them to what is safe to eat, and teaches them to drink water.

LAMBS

Particularly maternal does will also raise lambs, but this isn't generally encouraged because lambs bunt their mothers' udders much harder than kids do and it's easy for them to damage a doe's more delicate udder. It's also frustrating for does that try to lie out their wooly foster babies because sheep are a following species and will do their best to follow their foster mothers.

THE STOMACH IN KIDS & ADULTS

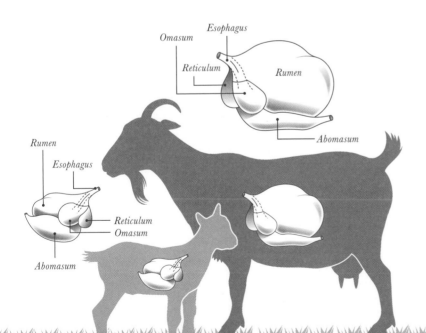

The esophageal groove is a reflexive tube running from the esophagus into the abomasum, stimulated by suckling (see also page 39). When a kid is small, this reflex allows milk to bypass digestion in the reticulo-rumen and go directly to the abomasum As the kid gets older and starts consuming roughage (at around 3 weeks of age), the other parts of the stomach start to grow in size and digestion begins to occur in the rumen. By adulthood, roughage is its main source of food.

Left: *In a kid, the abomasum accounts for 70 percent of the whole stomach. By the time a goat has reached adulthood, the abomasum has become much smaller, and the rumen much larger.*

HORNS

Depending on sex and breed, horn buds begin emerging anywhere from a week to 3 weeks of age. Kids scrub their foreheads on hard surfaces to help their horns break through. As horns begin growing, their outer surface sometimes seems to peel. That's normal. As the horns grow longer, they smooth out again.

Above and below:
Disbudding seems gruesome but it's over with very quickly when performed by a knowledgeable veterinarian or an experienced breeder. Kids recover in a flash and instantly race back to their mothers for reassurance.

Kid is placed in disbudding box, which allows only its head to protrude

The hair around the horn buds is trimmed, then a dehorner is applied for up to five seconds

The area is treated with an aerosol dressing before the kid is released to its dam

BUCKLINGS

When a buckling is born, his urethral process and the glans of his penis are attached to the inside of his prepuce (his sheath) by a frenulum membrane. As his body begins producing testosterone, sometimes when he's only a few weeks old, he'll begin practice-breeding his dam and sisters and may engage in a bizarre form of air-humping behavior in which he semi-squats and then repeatedly thrusts his hips. These sessions help break down the adhesion and allow him to extend his penis. When that happens, he's probably capable of impregnating females. Occasionally doelings engage in this behavior too, though the reason they do it is unknown.

KIDS' PLAY

Kids are agile and active and they begin leaping and running within a few hours of birth. Play behavior then increases in duration and complexity (mounting, butting, climbing) for roughly two weeks, and then continues into young adulthood. By early adolescence, youngsters have established their niche in their herd's hierarchy. Their physical capabilities and activity levels are almost as high as those of younger kids.

Initially, kids stay close to their dam, playing with siblings but usually not with other peers. Then at around 5 weeks of age they begin forming "gangs" with other kids and spending more time away from their dam, exploring, racing, play-fighting, tasting everything, and climbing on things, and only returning to their dam to nurse and sleep. Bucklings tend to play rougher than doelings and may attempt to play roughly with human handlers (see pages 110–11).

Below: Most does are eternally patient, allowing their kids to climb on them at will. Other does' kids, however, are immediately and violently rebuffed.

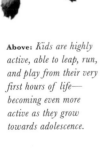

Above: *Kids are highly active, able to leap, run, and play from their very first hours of life— becoming even more active as they grow towards adolescence.*

Vocalization

The sound made by a goat is known as "bleating" or "calling." Goats are, as a whole, extremely vocal animals—even dozing goats will murmur in their sleep. A few may be silent types but they are rare indeed.

GOAT COMMUNICATION

Some does in heat, especially if they're penned away from their chosen buck, scream and moan in a loud, insistent manner, although others vocalize softly, if at all. A day to a few hours before kidding, does begin murmuring to their unborn kids in a low, sweet rumble that some call "mama's voice." They continue using it for a few hours to a day after their kids' birth but at no other time.

Kids begin mewling soon after they are born. Their mothers recognize them by sight and sound within four hours of birth, and within a day or so by sound alone—one study demonstrated that does remember their kids recorded cries for up to 13 months after weaning. Kids call gently to their mother and siblings but they will shriek loudly if picked up, separated from their dam, or otherwise badly frightened. When kids are handled for unpleasant procedures, for example, they let out a high-pitched distress call. Some kids vocalize more than others, too, calling to their dams and siblings, and later to unrelated playmates.

Bucks in rut will blubber—a vocalization that sounds like they're saying *wup-wup-wup*. They also "buck scream," a loud cry that is made in conjunction with stomping a foreleg while courting does (see page 83).

Goats in a group quietly and calmly call to one another to locate herdmates and they greet their caretakers with these same soft calls—unless it's feeding time, when they call more stridently. Goats in isolation, however, will scream loudly. And no one who has heard a goat scream in pain will likely forget it.

Right: Consider whether a breed is noted for vocalizing when selecting goats to keep in towns.

Below left and right: Swiss dairy breeds can produce considerable vibrato, while some American Pygmys almost bellow.

GOATS DON'T ALL SOUND THE SAME

Each goat has its own distinctive sound, and people who work with goats soon come to recognize the different voices in a herd. Breed makes a difference, too. Some breeds have considerable vibrato in their voices while others—Nubians and Boer meat goats, for example—have none at all. Some breeds vocalize more frequently and stridently than others, making them poor choices for goats kept in urban and suburban situations.

In 2019, however, a study of Pygmy goats determined that beyond the role played by genetics, groups of goats establish shared sounds. The study groups' calls were recorded at 1 week old, at which point full siblings' voices were the most alike, although half-siblings raised in the same social group were similar. However, after 5 weeks, all the kids in one group sounded much like other kids in the same social group. They'd developed group accents.

THE ALARM SNORT

When a goat is alarmed, it stands rigidly with all four legs braced, its head held high, and its ears pointed forward, and issues a distinctive, sneeze-like alarm call, intended to alert other goats to danger. This call is so much like a loud human sneeze that sneezing humans tend to startle goats.

Are Goats Smart? ✎

Cognition is the ability to think, thereby determining the degree to which a human or animal can perceive, learn, pay attention, problem-solve, and remember. So are goats smart? Yes, indeed. Every goat owner knows it's so but science is now proving that it's a fact.

STUDIES INTO GOAT COGNITION AND MEMORY

In a 2014 study, researchers from Queen Mary University of London conducted experiments in which they observed 12 goats of various breeds learning how to retrieve a food reward. To reach the reward, the goats needed to pull out a lever attached to a rope using their teeth or lips, then lift the lever to allow the food to drop from a dispenser into a feeding bowl. Most mastered the task in 12 trials or less. The goats were then retested 26 to 33 days after the initial experiment, and again 281 to 311 days later. All of the goats immediately manipulated the dispenser to claim their rewards.

In another study, 20 goats were shown pairs of black-and-white images mounted roughly side by side on a fence. Each image showed an unfamiliar person, either smiling or frowning. In four trials, 51 percent of the goats approached and

Below: *One of the 12 goats used for the Queen Mary University study carrying out the two-step task required to release a food reward, as photographed by study participant Elodie Mandel-Briefer.*

Side view of the feeding box

Pulling the lever

Lifting the lever

Claiming the food reward

the negative was shown compared to the positive, indicating greater interest in negative faces, perhaps because they found them threatening or disturbing.

RELATING TO HUMANS

Studies have also revealed interesting insights into the way goats rely on their interaction with humans when completing tasks. In one example, researchers taught 34 goats to lift the lid on a transparent plastic box to get a treat. Then they sealed it shut. The goats could see the food but not get it. Most of the goats examined the box then repeatedly looked up into the eyes of a nearby human and then back at the box as though asking for help. If researchers turned their backs, the goats didn't look up at them for as long or as often. And in a separate study, goats quickly learned to go around a barrier to find food after watching humans perform the task first. (For full details of all the studies mentioned here, see page 218).

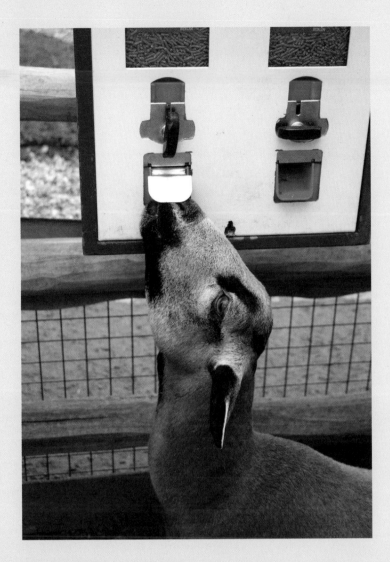

Above: *Tests like those mentioned on these pages have proved the reasoning ability, curiosity, and memory of goats beyond a doubt.*

nudged the happy face first; in 30 percent of the trials, they approached the angry one first, and in 20 percent, they ignored both faces. This was remarkable because in previous studies only dogs and horses demonstrated an ability to differentiate between human expressions.

In a similar study, 32 dairy goats were shown images of four familiar herdmates displaying positive and negative expressions. The goats spent more time with their ears forward when

GOATS VERSUS SHEEP

In 2013, researchers conducted a study comparing the short-term memory and reasoning ability of six East Friesian sheep and 12 Nigerian Dwarf goats by concealing a food reward under one of two bowls and letting them choose which one contained the treat. The goats outperformed the sheep at every level. (See also pages 64–65 for more on goats versus sheep.)

CHAPTER 4

Goat Management

Caring for Hooves ✐

Wild goats keep their hooves worn
down by climbing on rocks, as is true of
domestic goats allowed to roam on hard
or rocky surfaces. However, most domestic
goats must have their hooves trimmed to
keep them in the pink. How often this is
done varies, depending on conditions, but
even goats that live on rocky ground may
need their hooves trimmed several times
a year—and some may even require a
trim as often as every 8 weeks.

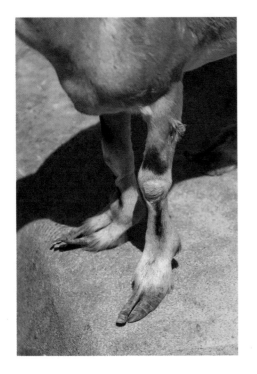

OVERGROWN HOOVES

Overgrown hoof walls, as seen in the
photo to the left, grow down and curl
under the sides and over the sole of the
hoof, eventually deforming the hoof and
crippling the goat.

Above: *Without sound,
solid, and shapely hooves
a goat cannot move to
pasture to eat, he cannot
climb, and he's far more
likely to suffer from
hoof disease.*

TRIMMING A GOAT'S HOOVES

Hooves are best trimmed with the goat standing, so that you can work on one raised leg at a time. Goats can be taught to stand quietly without leaning on you or falling over while you work. If hooves are exceptionally hard, you will find that trimming after a soaking rain helps immensely.

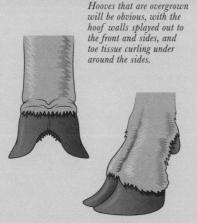

Hooves that are overgrown will be obvious, with the hoof walls splayed out to the front and sides, and toe tissue curling under around the sides.

1

You will need some handheld hoof trimmers and a simple metal rasp. Hoof trimmers designed for the purpose work best, but sharp garden pruners are acceptable, too.

2

Begin by using the closed tip of the hoof trimmers to pick out any dirt from every part of the bottom of the hoof.

3

The hoof walls will need to be trimmed until they're level with the bottom of the hoof—the base of the hoof should be parallel with the coronary band (see page 46). The best way to judge this is to use the parallel growth rings that run round the hoof as a guide.

4

Trim away all excess nail tissue around each toe, bringing the edges to the level of the soles in the center and getting rid of any ragged bits. Sometimes a simple touchup will be all that is required.

5

Finish off by using the rasp to level out the base of the hoof for a neat finish. Stop trimming when the feet look slightly pink to avoid any bleeding.

6

Kidding ✑

Kids are born after a roughly five-month gestation, varying from 142 to 155 days. Although a goat will give birth by itself, an understanding of pregnancy and birth, or kidding, is crucial for those occasions when assistance is needed.

Above: *Though does sometimes surprise their handlers by kidding seemingly without warning, there are usually plenty of signs to watch for in advance.*

SIGNS THAT KIDDING IS CLOSE

1. A first-time doe's udder will begin filling about a month before kidding; the udders of does that have previously kidded fill anywhere from a few weeks to immediately prior to delivery. Right before kidding, most does develop strutted udders—so engorged with milk that they become shiny— and their teats will stick out to the sides somewhat.

2. The pelvic ligaments in a doe's hindquarters attach at either side of the spine, midway between the hips and pin bones, then angle away toward the rear. To locate them, slide a hand along the doe's spine, including an inch or so to either side. The ligaments are the size of a pencil and very firm until secretion of the hormone relaxin gradually softens them in preparation for birth. When they can't be felt any longer, expect kids within 24 hours.

3. Relaxin will also cause the rump to become steeper as labor approaches, from hips to tail and from side to side. As this happens, the area along the spine seems to sink and the tail head rises. About 12 hours before kidding, it's possible to grasp the spine at the tail head and almost touch thumb to fingers on the other side.

4. The perineum, the hairless area around the vulva, sometimes bulges during the last month. About 24 hours before kidding occurs, the bulge diminishes and the vulva becomes longer, flatter, and increasingly flaccid.

5. As the cervix begins to dilate, the cervical seal (wax plug) liquefies, and mucus is discharged from the vulva (ranging from clear, thin goo, to a thicker, opaque white substance, to thick, amber-colored discharge). This can happen days or hours before a doe delivers.

FIRST-STAGE LABOR

Pawing out a nest in the pen

An early contraction (ears extended)

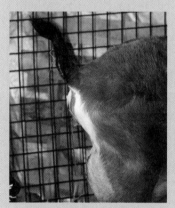
Stringing pre-kidding mucus appears

Harder contraction prior to full labor

Ears indicating pain and distress

Looking back at sides

During first-stage labor, relatively mild uterine contractions cause a doe to pause for a moment to stretch and raise her tail. She might also:

• Engage in nesting behavior, digging a depression in her pen, lying down, getting up, circling, digging, and repeating the cycle over and over.

• Search for something while murmuring in a low-pitched, soft voice.

• Wobble on her hind legs as hormonal changes relax her pelvis and affect her gait.

• Yawn and stretch. Stretching helps to put her kids into a proper birthing position.

• Drift away from the herd to seek a nesting spot, sometimes with her dam, a daughter or sister, or a best friend for company.

• Gaze into the distance with an unblinking, faraway look on her face.

• Pant, taking short, shallow breaths.

• Grind her teeth.

• Look or bite at her sides.

• Become unusually affectionate or standoffish.

HARD LABOR

Within 12 to 36 hours, second-stage labor begins. The doe lies down and rolls onto her side when a contraction begins. She rides out the contraction, and then rises and repositions herself until she finds a position she likes. She may roll up onto her sternum between contractions, but she'll usually remain lying down (although some does deliver standing up or in a squatting position).

Within around 30 minutes a fluid-filled sac called the chorion will appear in her vulva—one of two separate sacs that enclose a developing fetus in the womb; the other, inner sac is the amnion. Either or both sacs can burst within the doe or externally as the kid is delivered.

In a normal front-feet-first delivery, a hoof appears inside the chorion or vulva, followed by another hoof, then the kid's nose, tucked close to its knees. Once the head is delivered, the rest of the kid quickly follows.

In a normal hind-feet-first delivery, two feet followed by hocks appear. Because the umbilical cord is pressed against the rim of the pelvis during this delivery, it's

KIDDING POSITIONS

Most births proceed without problems but it's important to keep a veterinarian's phone number on hand, or to study diagrams of problem births and how to correct them.

Normal

Front feet first   *Hind legs first*

Breech   *Head turned down*   *Two tangled kids*

wise to *gently* help a kid out once his hips appear. Otherwise, both are textbook deliveries and rarely require assistance.

Help is required if kids present in abnormal positions (butt or side first, head turned down, or two kids coming out together with limbs tangled). Experienced kidding attendants soon learn to correct these problems, or a veterinarian can assist.

Below: *Once the tips of both front hooves and a nose emerge, you can relax. Unless the kid is stuck, at this point catastrophe has been averted.*

THE CHORION EMERGES

The chorion appears around 30 minutes after hard labor begins.

A first hoof appears within the chorion, signaling a front-feet-first delivery.

The rest of the body follows the hooves, still within the chorion.

Because of anatomical differences, a kid can't be pulled with tackle the way farmers pull calves, and it's crucial to use lots of lubricant. The kidding attendant pulls by hand while the doe is having a contraction, pulling out and down in a gentle curve toward the doe's hocks.

POST-DELIVERY

Most kidding attendants quickly remove birthing fluids from a kid's nose by stripping their fingers along its face, then they place the kid in front of its mother. She'll immediately begin licking and nuzzling it to establish the maternal bond. At some point the doe may leave the newborn kid to deliver another kid. This is normal. After the subsequent kid is delivered, the attendant places both kids in front of her so they continue bonding.

After the last kid is born, the placenta—a fluid-filled sac on a cord—appears and hangs from the vulva. This can take up to 12 hours, but usually happens within one or two. If she doesn't pass it within 12 hours, veterinary assistance is needed. Indiscriminate pulling on a partially expelled placenta

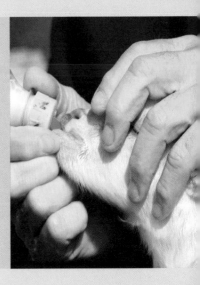

NURSING BY HAND

Does often reject sickly kids, so they will need feeding by hand. This should be done with a bottle rather than a pan, due to the structure of a kid's stomach (pages 39 and 89). Neonatal kids are also wired to seek sustenance in dark places like their dam's armpits or groin, and are attracted to warm, bare skin, so it helps to cover a kid's eyes when teaching it to nurse from a bottle.

can cause prolapse. There may be one placenta for the litter, one for each kid, or any other combination.

Does are wired to eat their placentas, possibly because cleaning up after themselves helps keep from attracting predators. While it's a natural thing, eating the membranes presents a choking hazard, so attendants usually collect and dispose of them. And a small amount of bleeding during kidding and up to a week or so after is of no concern—this is simply nature's way of cleaning out the uterus.

Below: The first kid should be born within an hour after hard labor commences, and a second kid will usually follow around 30 minutes later. An hour or two after the birth of the last kid, the placenta will appear, although this can sometimes take much longer. There may also be more than one placenta.

FULL BIRTH & THE PLACENTA

The first kid is born and is licked clean by the doe.

A second kid emerges around 30 minutes after the first.

The placenta appears around an hour after the last kid is born.

Common Health Issues

Goats are basically hardy creatures but they do require care to keep them in the pink. Bucks require careful feeding to prevent urolithiasis, all goats benefit from mineral supplementation, and internal parasites are a problem wherever goats are kept.

UROLITHIASIS

One of the most serious afflictions a male goat can experience is urolithiasis—more commonly known as urinary calculi or water belly: uroliths, or "stones," in the urinary tract. Stones form when urine pH levels cause certain minerals to bind and form crystals that gradually become solid. Such imbalances can occur in situations where dairy does are fed grain to boost milk production and males receive the same rations, when pets are fed a high-in-grain rations because owners don't know better, or when feeding for rapid weight gain in show or meat goats.

Below: Although stones can form in goats that eat a natural diet of browse and grass, it rarely happens. They usually form when animals are fed a diet rich in grains, most of which are high in phosphorus and low in calcium.

THE MALE ANATOMY

Stones can form in the female urinary tract but they pass easily and rarely cause problems. The male urinary tract, however, is more complicated. A male goat's penis has an S-shaped curve in it called the sigmoid flexure, and the urethra extends as a short slender structure (the urethral process, or pizzle) around an inch (2–3 cm) beyond this (see the diagram on page 81). Stones can become lodged in these structures, potentially causing uremic poisoning and eventually, bladder rupture and death.

Early signs of a blockage include straining to urinate, dribbling urination, and abdominal pain indicated by stretching out all four limbs, kicking at the abdomen, and twitching of the tail. Later signs are loss of appetite, lethargy, crying out, swelling around the prepuce (the skin sheath enclosing the penis), and crystals in hair around the penile opening. If the urethra has ruptured, the abdomen becomes distended, hence the term "water belly." Eventually the goat will lie on his side and refuse to rise. Then he'll have seizures and die.

PREVENTION IS BETTER THAN A CURE

Since surgery may be required, prevention is the best solution.

• Supply rations with a ratio of 2 parts calcium to 1 part phosphorus. When feeding grain, choose bagged rations formulated for sheep or goats, not for other species.

• Male goats fed high-grain rations can have salt added to the mix at the rate of 2 to 5 percent to increase urinary output (and avoid urine concentration).

• Ammonium chloride is another useful dietary addition, readily found in rations formulated for goats. It makes any crystals in the urinary tract more soluble.

• Dehydration contributes to stone formation, so provide access to clean, palatable water at all times.

• A male goat's penis stops maturing when he's castrated, but it's easier for stones to pass through a full-size penis, so most US veterinarians suggest delaying castration as long as possible.

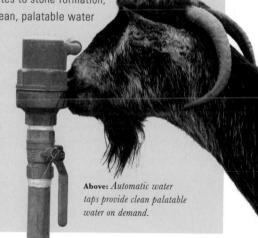

Above: *Automatic water taps provide clean palatable water on demand.*

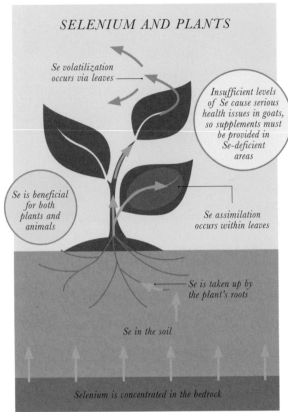

SELENIUM AND PLANTS

Se volatilization occurs via leaves

Insufficient levels of Se cause serious health issues in goats, so supplements must be provided in Se-deficient areas

Se is beneficial for both plants and animals

Se assimilation occurs within leaves

Se is taken up by the plant's roots

Se in the soil

Selenium is concentrated in the bedrock

MINERAL SUPPLEMENTS

Goats should be provided with loose minerals or mineral tubs containing minerals deficient in the soil of areas where they live or where their feed is raised. This varies from place to place but the major culprits are cobalt (deficiencies lead to loss of appetite, emaciation, weakness, anemia, and decreased production), copper (symptoms of deficiency include anemia, dull and rough or bleached-looking hair coats, diarrhea, weight loss, and atrophied muscles), and especially, selenium (Se).

Soils in much of the northeastern and northwestern US and adjacent Canada, the southeastern US, and the UK (especially northeast Scotland and parts of southeast England), are selenium-deficient, as are parts of Germany, Denmark, Scotland, Finland, and most of the Balkan countries. Seriously selenium-deficient goats suffer from a malady called white muscle disease (WMD) and may suffer poor conception rates, stillbirths and miscarriages, retained placentas, or deliver weak kids. Kids born with WMD are weak and may be unable to suckle or stand. They can be treated with selenium injections but a better ploy is to treat adults for deficiencies before breeding. Does should be supplemented with selenium in their diet, and then given selenium–vitamin E (BoSe) shots about a month before kidding is due to occur.

Above left: *Goats need access to goat-specific minerals. Loose mixes served from waterproof feeders are the norm but minerals blended with solidified molasses and fed from tubs work, too.*

Above: *The micronutrient selenium is crucial for humans and animals, and plants provide the main dietary source. Selenium in the soil is absorbed by plants through their roots and then released into the atmosphere as vapor.*

NEMATODES

An overabundance of nematodes ("worms") impacts goats' health and can even kill them. Species vary from place to place, but one nematode that plagues goats on an international scale is *Haemonchus contortus*—the barber pole worm. Found in Asia (Indonesia, India), the UK, Europe (the Netherlands, Italy), South America (Brazil), Africa, Australia, New Zealand, and the US, it's adapted to conditions ranging from cold, mountainous regions to the tropics. It is, however, more prevalent in hot, humid regions than cold, dry ones.

Barber pole worms measure around 3/4 to 1 in. (18–30 mm) and live in the abomasums of ruminant animals (see page 38), where they suck blood, causing serious anemia and progressive weight loss. Because they're somewhat transparent and full of blood, they appear to be red. Females have white ovaries that wind around their bodies, hence the nickname "barber pole."

If you suspect an infestation, take a fecal sample to a veterinarian for testing. They will be able to distinguish between the presence of barber pole worms, liver flukes, lungworm, and others. An "in the field" approach for barber pole worms is to check goats' mucus membranes for signs of anemia. Gently pulling the goat's lower eyelid away from its face and examining the tissue inside the eyelid. It should be deep pink; if white, the animal is severely anemic and could die. This process is called FAMACHA© testing. A FAMACHA© chart provides five different colors, ranging from an optimal deep pink (1) up to a fatal white (5); comparing an animal's eye to the colors on the chart is

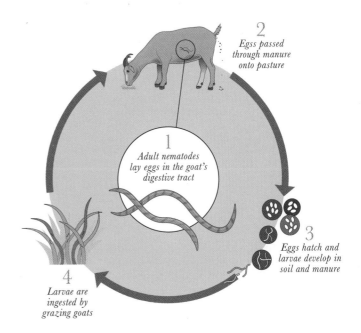

2 *Egss passed through manure onto pasture*

1 *Adult nematodes lay eggs in the goat's digestive tract*

3 *Eggs hatch and larvae develop in soil and manure*

4 *Larvae are ingested by grazing goats*

an easy way to establish whether it remains within acceptable levels, or whether it needs treatment. Barber pole worm and other nematodes are becoming increasingly resistant to the anthelmintic drugs ("dewormers") used to treat them, so dosing advice from a veterinarian or agricultural specialist is needed.

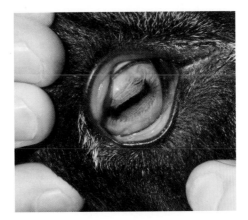

Above: *Goats ingest the larvae of barber pole worms when grazing, then release eggs through their manure, thereby facilitating the worms' life cycle. It is therefore crucial to deworm goats with an appropriate anthelmintic as needed, well before large populations wreak havoc on their hosts.*

Left: *Checking the color of a goat's lower eyelid will give an indication of whether a goat needs treatment for barber pole worm. The membrane should ideally be a dark, deep pink.*

An Unlikely Predator

Goats are easily injured or killed by predators. They're prey animals and essentially defenseless (horned goats might look intimidating but even they are wired to run rather than stand their ground). When one thinks of predators, wolves, bears, coyotes, and mountain lions spring to mind. However, the most common killers of goats worldwide are actually dogs—sometimes feral dogs but more commonly, family pets.

THE THREAT OF FREE-ROAMING DOGS

A recent study in Australia tracked 1,400 dogs that had attacked livestock and learned that most of them were backyard pets—dogs ranging from three months to 12 years old, both purebreds and mixed breeds. Even dogs of breeds bred to not harm other animals apparently could not resist the urge to attack livestock. The study also showed that, contrary to what dog owners often think, dogs do not necessarily attack in packs; almost all of the recorded attacks were by single dogs, or pairs.

Goats, sheep, and poultry bear the brunt of dog predation. These animals flee when dogs attack, and this is the type of action that feeds predatory behavior.

Dogs chase prey for fun, and will climb or dig under fences to reach their target. While other predators generally kill one animal every few days and then eat them, neighborhood dogs may kill several animals in a single night. They will often pursue an entire herd at top speed, and they usually go for a goat's hind legs and rear end, although they will also rip off ears and faces, peel off strips of hide, and generally mutilate their prey without killing them outright. The goats die later from injuries or exhaustion—or end up having to be euthanized—and pregnant does will often abort their kids.

Below: *Goats that are kept in wide open spaces, like this outback farm in Australia, are vulnerable to attack by dogs.*

PREVENTATIVE MEASURES

Free-roaming dogs, be they feral dogs or domestic pets, are generally present wherever goats are kept, so to keep them safe, goat keepers must take precautions at all times, starting with good fencing.

Tethered goats are at the mercy of dogs and other predators, so it's best not to tether them at all. However, when it's necessary, it should be within sight and sound of help. If the goats need to be moved around, portable electrified fencing is better than tethering; electric fencing keeps goats in and helps keep dogs at bay. (Goats will sometimes need to be trained what an electric fence is or it will not work.)

Most dog attacks happen at night or early morning, so it's wise to lock goats in secure facilities overnight. Motion- activated lights in the barnyard help, too.

Left: *Anatolian guardian dogs were first developed in Turkey.*

LIVESTOCK GUARDIAN DOGS

A highly effective approach to dog predation is to keep a livestock guardian animal with the herd. Llamas and donkeys are used in some locations but worldwide, livestock guardian dogs are the first choice. These are large, strong dogs, bred specifically for the task: to bond with their charges and protect them from harm. Such dogs may need to be raised with the goats, though—there may potentially be issues if a dog does not consider its goat charges to be family.

Developed in Western and Eastern Europe largely to protect sheep and goats from wolves, popular livestock guardian dogs include the following breeds:
• Great Pyrenees and Pyrenean Mastiff of France and Spain
• Maremma Sheepdog of Italy
• Kuvasz and Komondor of Hungary
• Karakachan of Bulgaria
• Tatra of Poland
• Gampr of Armenia
• Akbash, Anatolian Shepherd, and Kangal dogs of Turkey
In the United States, mixes of these guardian breeds are popular.

Goats Behaving Badly ∽

There is one bad behavior nearly every goat keeper encounters sooner or later: aggression. This is the result of goats' wired-in need to establish their place in their herd's hierarchy. It's often directed at caretakers, whom aggressive goats perceive as lower-ranking members of the herd, and the culprits are usually bucks, particularly during rut (see page 80).

SIGNS OF AGRESSION

Aggression often begins early. People allow kids to head-butt and jump on them because it's cute. But it isn't when the goat weighs 200 lb. (90 kg). The first signs of true aggression may surface when a goat is between one and two years old and ready to establish its ranking in the herd. It may ignore you, rush a gate as you pass through, or become obnoxious while being fed. If these behaviors go uncorrected (getting angry doesn't count), it may begin to posture, with head held high and tipped forward, commonly used when challenging other goats. A posturing goat will move closer and closer and eventually turn sideways, blocking your path—a behavior that no goat would tolerate from a subordinate.

Next the goat will issue a physical challenge by rearing up and swooping down. It might "accidentally" slam into you, or swing its head in a challenging manner. Finally it will bite, hook its horns at you, or head-butt you.

Above: Although it can be fun to play at head-butting with a goat while they are young, aggressive play like this should never be reinforced.

Right: Goats challenge humans in the same way as they challenge each other. Pushing, shoving, head-butting, biting, or standing with their front feet on your chest—all must be strongly discouraged.

FEEDING TIME

When space at the feeding station or water trough is limited, high-ranking goats sometimes guard resources from low-ranking individuals, so multiple feeders and watering stations with enough per capita space to serve all herd members at the same time are essential.

Right: Goats hate getting wet. Therefore a squirt of water in the face provides an effective yet harmless deterrent when goats crowd or jump on humans.

BECOME A HIGHER-RANKING "GOAT"

The solution is to assert superiority early on. When a kid jumps up, place it back on the ground, but do not push on its forehead, which encourages it to push back. Teaching a goat patience by securing it to a fence for grooming and hoof work is another good idea. And goats should be taught to give way when you enter the paddock or pen, even if you are carrying feed. Goats hate being wet, so a blast from a water hose does a good job of reinforcing the demand. If cornered by an aggressive buck, grab his beard and use it to lead him to a barrier where it's safe to let him go.

Incorrigible goats that pose a danger to handlers should be thrown. To do this the goat should be wearing a collar or halter for control. Hold the goat's collar with one hand, lean across its back, and grasp the opposite hind leg. Now throw your body into the goat's side while pulling up and back, so that the goat lands on its side. Immediately throw your body over the goat's, with an elbow against the

AGGRESSIVE GOATS AND CHILDREN

Goats that never challenge adults sometimes head-butt or knock into small children, probably perceiving them as lesser members of the herd, so keep toddlers out of goat pens and paddocks, or give them something to raise their status. A squirt bottle usually works nicely, or a short, sharp blast from a canned air horn, to be used only if threatened.

goat's neck so that it can't raise its head, and hold the goat down until it stops struggling, then an additional five minutes or so. If you are not tall enough to reach across the goat, reach under it and grasp the hind leg and front leg farthest from you and pull. While this may sound extreme, it usually only has to be done once and might keep an unsafe goat from going to slaughter or at least being sold.

Above: Vigilance is a must when small children interact with goats. Goats that would never shove or head-butt an adult sometimes perceive toddlers as lower-ranking herd members.

Handling & Training Goats ✒

Goats can be taught to perform complex tasks, as has been proven in various studies (see pages 94–95), and they also learn everyday tasks with ease. But understanding just how goats perceive their world is the key to handling them with ease and a minimum of stress.

The trick to handling and training is to do it with kindness. Goats are intelligent and have long memories for good or for bad. A few quiet words, a gentle stroke of the neck, or a food tidbit go farther to earn their respect and cooperation than any amount of shoving, shouting, or striking.

Below: *A group of Valais Blackneck goats stick together in their Alpine home. Understanding how goats operate naturally as a group makes it much easier to work with them rather than against them.*

USEFUL TIPS FOR TRAINING GOATS

• Calm goats generally move in family groups, with the oldest females and their extended families at the head of the pack. Some goat handlers bell these leaders to facilitate group cohesion.

• Because goats lack a flocking instinct, it is much easier to lead them—perhaps with a bucket of feed—than to drive them. (Goats can be herded—although this is usually done along with sheep, with their greater flocking instinct—but leading them is much easier.)

• Goats move when other goats move. Lead the herd queen and the rest will come along.

• Goats hate to be rushed. Move them only as quickly as the slowest goat in the group.

• Goats readily move into larger areas because they don't like to be tightly enclosed. They also prefer to move on the flat and up an incline rather than down it.

• Goats move more readily into the wind instead of downwind. They would also rather move toward light than into dark or uneven lighting, and they move more efficiently toward an opening than to a dead end. They resist being driven through narrow openings, so open doors and gates as wide as they'll go.

• Goats like routine and move best if the same familiar paths and flow directions are used every time. Lanes leading to pastures and handling yards allow goats to move freely rather than being driven.

• Goats react negatively to loud noises, sudden movements, and excessive force. Terrified goats pack into corners where those closest to barriers can suffocate or be injured. They are easier to handle when separated into compatible groups in smaller enclosures rather than a large mob in one big enclosure. Be patient. Give goats time to assess a situation. Goats remember negative experiences for a very long time.

• Kids and young goats move through facilities more efficiently when moved with mature, well-trained adults. Suckling kids that become separated from their dams try to return to the place where they last saw her.

• Goats strenuously object to being handled by their horns. If this becomes necessary, grasp them close to the goat's head instead of near the tips; mishandled horns can be broken during handling, creating a painful and bloody mess.

• If a goat stops, it's counterproductive to push it. Pull it sideways until it starts moving again, or lightly tweak its tail up and forward.

• Mixing groups of goats in handling yards upsets each group's social hierarchy. It's best to handle one bonded group at a time.

• It's much easier to keep track of a goat fitted with a collar and bell.

Above and below: *Clicker training is employed mostly by owners of pet goats, although it is also a popular technique with those who work with packgoats (see pages 136–37).*

CLICKER TRAINING

Goats are an inquisitive, intelligent species that loves food, so they respond best to positive training involving praise and food rewards rather than methods that punish mistakes, and one of the best methods for this is clicker training—commonly used to train dogs, and easily adapted to use with goats.

The clicker is a small, hand-held device that is used to make a sound at the precise moment that the animal performs a desired action. This clear signal is then backed up by a tidbit of food, building a positive association that increases the probability that the animal will repeat the behavior next time it's asked. Using the clicker, goats can be taught everything, from behaving for procedures like hoof trimming and deworming, to pulling a cart or performing tricks.

AFFECTION TOWARD HUMANS

Goats show affection by gazing into the faces of humans they like. All goats have scent glands in close approximation to their horns, or where their horns would be, so they rub their foreheads on people they care for to distribute their scent (which only goats can smell). Some friendly bucks will also attempt to urinate on humans to show affection, as they would a doe—and, even worse, this often happens during rut, when there lots of pheromones are present (see pages 80–81)!

HOUSE GOATS

Goats have always been kept as farmyard and backyard pets. But because they're easily house trained if started young, an increasing number of goat owners are now also keeping them as house pets.

Keeping goats singly is generally not recommended—goats need companionship—but the exception are house goats, which are raised as bottle kids to bond with humans. Sometimes kids raised in this way will eventually move out with a main herd, but keeping a house goat is a personal choice, and many house goats live very happily as single goats all their lives.

Training a kid not to urinate indoors is much like training a puppy. Kids are wired not to urinate where they sleep, so it's important to keep them in a close enclosure, such as a large wire crate or a children's playpen, complete with bedding—fleece throws, towels, or pieces of blanket work well. When a kid is out of its quarters and preparing to urinate, scoop it up and carry it to a designated spot. When the kid finishes urinating, praise it profusely.

This must be repeated after napping, meals, play, and every few hours through the night, although by around 3 weeks a single nightly trip is usually sufficient. Provisions should be made for weather, too—even tiny goats hate to get wet. If accidents happen in the kid's enclosure, replace the bedding and clean up with a dilute bleach solution, leaving no odors to tell the kid it's okay to relieve itself there.

Most young kids also poop at their urination spot, though their ability to hold back feces for long periods actually lessens as they grow older.

HOUSE-TRAINING TIPS

• Cover a kid's crate with a blanket at night; this sends a signal that it's time to sleep.

• Hang a few bells in the crate. These will alert you if the kid is moving around at night and needs to urinate.

• If you can't be at home to take the kid to its spot in a timely manner, it's best to have an alternate place outside in a garage or barn where it's alright for the kid to urinate while you are away. This is also a workable alternative if you don't want to get up in the night. Keep in mind that goats are social creatures, so if a kid is kept outside and away from humans for any length of time, it needs a companion.

• When training to urinate outdoors, it's easy to train more than one kid at a time, especially if they share a sleeping enclosure. If training to go indoors, a large urination station like a children's wading pool is essential.

Above: *Indoors, puppy housebreaking pads, a thick, launderable blanket, or a small wading pool or tray filled with sand or cat litter are good choices.*

Goat Play ✒

Goats begin playing when they're a few hours old and, body willing, they play until they die, leaping, twirling, sprinting, feinting, and climbing with impunity. Because they're so playful by nature, it's good to give them room to run and things to play with, and it's essential for goats confined indoors.

BEHAVIORAL ENRICHMENT

Numerous studies have shown that providing animals (including goats) with a stimulating environment will elicit natural behaviors from them, including exploration, foraging, locomotion, and social interaction—an approach known as behavioral enrichment.

In one study, Brazilian researchers separated 12 Saanen dairy does into two groups, placing one group in a barren enclosure and the other in an enclosure containing enrichment materials (bottles filled with ground corn, brushes fixed to the walls for self-grooming, a suspended tire, and a tree trunk for climbing).

Right: *Goats form fast friendships even as kids, and together they will climb on everything they can, including patios and decks, automobiles, picnic tables, and lawn mowers.*

Above: *Goats thrive on climbing, descended as they are from mountain goats. Satisfy their urges by building them a sturdy playground. Even a downed tree will do.*

Right: *All sorts of items make perfect climbing toys. Trampolines are favorites with some herds, as are wooden cable spools from the electric co-op, or easily turned-over tubs.*

The goats in the enriched environment spent more time self-cleaning, eating, and interacting socially, and less time standing around or resting. They were also observed to be much more physically active.

Although frequently used for captive animals, pets and farm animals can benefit from species-specific enrichment strategies. Enrichment is especially important for goats because if they don't have something to do they'll create their own entertainment, which can range from purposely destroying fences to beating up their herdmates and caretakers.

ENRICHMENT PLOYS
FOR GOATS

As browsers, goats use their mouths to investigate their world; they are also excellent climbers. Any toys that exploit these characteristics work very well indeed. Goats also like variety, so toys should be moved on a weekly or biweekly basis to stimulate investigative behavior.

The best toys are made of sturdy metal, hard rubber, or unbreakable plastics, and these can be placed on the ground or suspended from ceilings, walls, doors, or fences where goats can mouth them and use them to make noise.

It's important to provide enough toys so that all goats have a chance to use them at the same time. Otherwise goats high in the herd hierarchy (see pages 76–79) will guard them and lower-ranking goats will be left frustrated.

Goats should be given nothing that has previously contained toxic chemicals or that has holes big enough for a leg to slip through and lead to injury. Eschew any items with sharp edges, as well as those that can easily be chewed apart and then swallowed. Any play items should also be checked regularly for exposed nails, weak or rotting spots, splinters, breakage, and holes.

It's also worth remembering that additional human contact is an enrichment ploy in itself. Being taken for walks, groomed, and taught new tricks are both fun and stimulating for goats.

Above: *Time out and about is perfect fun for a pet goat. Teach him to lead and to hop up into the back of your car, then off you go!*

Left: *Children's plastic playhouses provide a stimulating structure to interract with. Being designed with children's safety in mind, these playhouses are also free of sharp edges, splinters, and nails.*

SUGGESTIONS FOR TOYS

Suspend a ball in their stall, paddock, or pasture, or try sturdy balls designed for horses to play with. Some of these have a handle to make them easier to grasp.

Commercially made dog and horse toys that dispense treats as they're played with work well with goats, as do plastic pitchers and water cooler bottles with treats inside (with the caps left off, or small holes cut in the sides).

Hay bales, logs, plastic cattle mineral tubs, wooden tables or benches (with legs partially buried in the ground so they can't tip over), tractor tires (upright and partially buried), wooden packing crates, and empty electrical wire spools can all provide an inexpensive climbing apparatus. Or use discarded wooden pallets to make a multi-level structure.

Children's plastic playhouses, slides, teeter-totters, and specialty beds made to resemble cars, trains, and planes.

Tires suspended from trees. Leave it on the rim for extra weight if hanging for bucks.

Full-size trampolines provide lots of fun and mini-trampolines (rebounders) make comfortable beds. Secure large trampolines so they won't tip over, and provide a jump-up box for small goats.

Plastic scrub brushes attached to walls and fences can be used for self-grooming.

Large, sturdy, leafy tree limbs or saplings provide climbing opportunities and green leaves to browse on, but positively identify the tree to make certain it isn't poisonous to goats (see page 135).

Below: *Scrub brushes are fun to shove around and perfect for scratching itchy, hard-to-reach places. Municipal utility departments sometimes give away worn-out street cleaning brushes. Ask!*

CHAPTER 5

Goats & People

Goats for Milk ✎

The genetic mutation that enables adult humans to digest milk occurred somewhere in the area between central Europe and the central Balkans about 7,500 years ago—a good thing, because goat milk is good food. Although cow's milk is by far the more common choice in places such as the United States and Europe, more than 65 percent of the world drinks goat's milk rather than cow's milk (although in terms of volume, this does not rival cow's milk—goat's milk accounts for around 2.4 percent of world milk consumption).

MILK PRODUCTION

Goats are often called the poor man's cow. They cost less to buy and maintain than most dairy animals, they produce nourishing milk in ideal amounts for immediate household consumption, and they thrive on native browse and in unfavorable climatic conditions.

The major milk-producing countries are found in the Mediterranean region, South Asia, and parts of Latin America and Africa, although the average milk yields vary significantly between these. In Bangladesh, for example, the average goat milk yield is about 176 lb. (80 kg) a year, while in India and Pakistan this figure rises to more than 309 lb. (140 kg). In turn, as a result of improved genetics and feedstuffs, typical dairy goats in Europe and North America produce considerably more milk per year than those in developing countries (see box).

Below: *Pashmina goats being milked in a Ladakh village in India. These goats are also raised for their fiber (see page 132).*

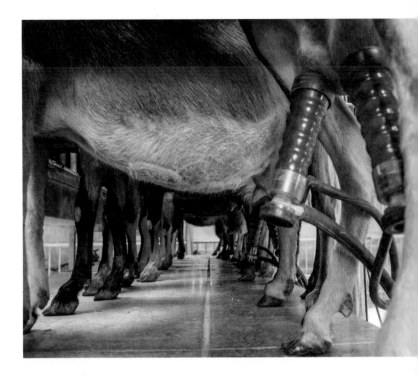

Typically, dairy does are bred once a year in the fall. When they "freshen" (give birth to kids and come into milk) they're milked for 10 months, taking a 2-month break to revitalize before kidding again. Production peaks 4 to 6 weeks after kidding and tapers off gradually in the fall. An average doe's annual production peaks during its second or third lactation and begins decreasing at 6 or 7 years of age, but many produce well for 10 years or more.

Some but not all does from high-production breeds can be "milked through," meaning they can be milked year round without being rebred. Milking through provides a year-round supply of milk without having to deal with rearing and selling kids. Production tends to diminish as time passes but lactations of 4 or 5 years are not uncommon.

The milk fat content of goat's milk varies from roughly 3 to 6 percent. It varies from goat to goat depending on breed, genetics, the types of feed consumed, and the stage of lactation. Typically, high-production breeds like Saanens produce lower-butterfat milk, while lower-production breeds like Nubians and Guernseys give higher butterfat counts. Some tiny dairy goats like Nigerian Dwarfs produce a surprising amount of milk with a high butterfat content (see page 68), though rarely as much quantity as full-size breeds.

A COMPARISON OF DAIRY BREEDS

Breed	Milk produced (in a 275–305-day period)
Alpine	2,620 lb. (1,188 kg)
LaMancha	2,349 lb. (1,065 kg)
Nigerian Dwarf	813 lb. (369 kg)
Nubian	1,963 lb. (890 kg)
Oberhasli	2,101 lb. (953 kg)
Saanen	2,765 lb. (1,254 kg)
Sable	2,574 lb. (1,168 kg)
Toggenburg	2,232 lb. (1,012 kg)

Source: 2018 Dairy Goat Herd Improvement figures, from the American Dairy Goat Association

MILK COMPOSITION COMPARISON CHART (per 100 g)

SOURCE	ENERGY (kcal)	FAT (g)	CHOLESTEROL (mg)	PROTEIN (g)	CALCIUM (mg)
Human	70	4.38	14	1.03	32
Cow	61	3.34	14	3.29	119
Goat	69	4.14	11	3.56	134
Sheep	108	7	–	5.98	193
Water buffalo	97	6.89	19	3.75	169

Source: US Department of Agriculture, Composition of Foods, *Agriculture Handbook 8-1*

THE BENEFITS OF GOAT'S MILK

Goat's milk has many properties that make it similar to human milk, making it ideal for feeding babies, especially those intolerant to cow's milk, and a healthy alternative for adults. It contains less alpha S1 casein, the compound responsible for most allergies to cow's milk, but more oligosaccharides which reach the intestine undigested and act as probiotics. There is also less lactose (the natural sugar found in milk) and fewer fat globules in goat's milk than in cow's milk, making it easier to digest. It features more essential fatty acids, which provide quick energy but are not stored as body fat. And it's rich in calcium, phosphorus, magnesium, zinc, selenium, riboflavin, and vitamin A. It is, however, low in vitamin B6, B12, and vitamin D.

Below: *Goat's milk has long been a staple in parts of Asia, Africa, and Latin America, but it is slowly becoming a more popular choice in the West.*

EASY CHEESE

Paneer, or panir—also known as chhena or fonir—is the essence of easy-to-make, mild-tasting, soft goat cheese. Heat a half gallon (4 liters) of goat's milk in a stainless steel pot. Stir it occasionally over medium heat until the milk reaches 185°F (85°C). Remove from the stove and then stir in ¼ cup (60 ml) vinegar or lemon juice. Ladle the curds into a colander lined with cheesecloth (muslin), fold in fruit or herbs if desired, then tie the corners and hang the bundle to drain overnight. This makes a round of cheese measuring 1 in. (2.5 cm) by 4 in. (10 cm) .

Paneer cheese round

Globally, goat milk is also used for crafting yogurt, fermented beverages like kefir and quark, and hundreds of cheeses including international favorites like chèvre (France), caprino (Italy), geitost (Norway), halloumi (Cyprus), kefalotyri (Greece), majorero (Spain), and paneer (India). Most goat cheeses have a characteristic tangy flavor that comes from caprylic, capric, and caproic fatty acids in goat's milk.

However, fresh goat milk produced by healthy animals kept and milked under sanitary conditions, when properly handled and promptly chilled, is sweet and tasty, with no hint of the tang associated with goat's cheese (although some goats do produce slightly off-tasting milk, so it's best to sample a goat's milk before buying her). And because it lacks the pigments that color cow's milk, goat's milk is white.

GOATS AS WET NURSES

Goats were commonly used in the eighteenth and nineteenth century as wet nurses for orphaned infants. In 1775, a French doctor, Alphonse Leroy, described the use of goats in a foundling hospital in Aix-en-Provence: "The cribs are arranged in a large room in two ranks. Each goat that comes to feed enters bleating and goes to hunt the infant that has been given it, pushes back the covering with its horns and straddles the crib to give suck to the infant." The image shown here was taken in Cuba. Although this now seems a strange custom, goats offered distinct advantages. They offered large, easy-for-a-child-to-grasp teats, an abundant supply of healthy milk, they bonded closely with their human charges, and unlike many human wet nurses of the day, they didn't expose their charges to syphilis.

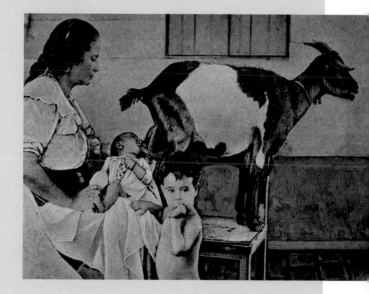

Goats for Meat

Most recognized breeds worldwide are kept for both meat and milk—Anglo-Nubian and Kinder goats are, among many others worldwide, dual-purpose milk-and-meat breeds that produce enough meat to warrant slaughtering. However, there are exceptions: goats that are bred purely for their meat.

MEAT GOATS

Although the flesh of any healthy goat can be eaten, some breeds produce more meat than others. The best known meat goat is the floppy-eared, Roman-nosed, powerfully built South African Boer goat (see page 183)—*Boer* in Dutch means

Below: *Most of the goats that are raised around the globe will end up as meat at the end of their productive lives but some are raised specifically for meat, like these young Boer goats.*

Above: Another leading meat breed is the hardy Kiko of New Zealand. Developed by crossing British dairy bucks with feral does, they are low-maintenance, high-output meat producers.

goat, South Africa also has the Savanna and the Kalahari Red, while elsewhere in Africa there is the Mashona of Zimbabwe, and the Galla meat goat of Kenya.

Another meat breed becoming immensely popular in the Americas and parts of Europe is New Zealand's Kiko goat, *kiko* being the Maori word for "meat" (see page 184). Development of these goats began in the 1970s when a group of large-scale farmers who were already farming enormous numbers of captured feral goats, formed a corporation known as the Goatex Group. The goal was to produce an extremely hardy meat goat capable of surviving without intervention under rugged conditions, while producing fast-growing meat kids. Foundation animals were not assisted at kidding, they did not receive supplementary feed or shelter, their hooves were not trimmed, and they were not dewormed. The weak and crippled died, and the goats that failed to perform were rigorously culled. All of today's Kiko goats descend in full from that group.

Some of the breeds that are raised specifically for meat elsewhere in the world include the German Fleischziege, or German Meat Goat, and the Bravia of Portugal, which are raised for the production of kid meat (see page 128). In Italy we find the Lombardy, and in Morocco the Moroccan Black. Chinese meat breeds include the Haimen, Banjiao, and Shaanan White, and Pakistan, too, is home to a number of breeds, among them the Teddy, Chappar, Lohri, Pamiri, and Tharki. In Thailand the Katjang is a major meat producer, as is the Terai goat of Nepal.

"farmer," and *boerbok*, "farmer's goat." The breed's exact origin is unclear, but today's Boer goats likely descended from indigenous African goats crossed with European dairy breeds by early Dutch settlers in South Africa. Modern Boer goats were then developed during the early 1900s, and in 1970 they became the first goats in the world to be scientifically performance-tested for meat production. Since then they've made their way around the globe, with sizeable populations in the United States, Canada, South America, Britain, Australia, New Zealand, and parts of Europe, where purebred bucks are often used on other breeds to produce meaty crossbreds for the commercial meat market. Today, in addition to the Boer

GOAT MEAT

Goat meat falls into roughly two categories: the meat from old, excess, or otherwise culled dairy or fiber goats, and meat from goats raised specifically for that purpose. The meat that comes from goats is either cabrito (or capretto)—mild, tender, light-colored meat from milk-fed kids—or chevon, the stronger-tasting flesh of older kids and mature goats.

A typical 12- to 20-week-old kid yields between 12 and 25 lb. (5.4–11.3 kg) of meat, while a goat that is 12 to 18 months old and weighs 100 lb. (45.3 kg)—will yield between 40 and 50 lb. (18–22.7 kg).

Goat meat is a healthy eating choice for many reasons: it's naturally lean, and much lower in saturated fat, cholesterol, calories, and sodium than beef, pork, chicken, or lamb.

Ribs, loins, and tenderloin goat meat are suitable for quick cooking, but for the most part, due to its low fat content and lack of marbling, goat meat can lose moisture and toughen dramatically if cooked at high temperatures, so it's best cooked slowly in stews and curries. Ground goat meat can also be successfully substituted for ground beef, pork, or poultry. Goat meat freezes well and can be canned.

Left: *All goats produce edible meat, though of lesser quality and in lower quantities than the large, muscular breeds specifically developed for meat production, such as the Kiko shown here.*

Right: *A herd of goats trek through the mountainous landscape of Yangshuo, near Guilin. China is currently the world's largest consumer of goat meat although its popularity is on the increase elsewhere.*

NUTRITION COMPARISON CHART (per 3 oz. /85 g cooked)

MEAT	CALORIES	FAT (g)	SATURATED FAT (g)	PROTEIN (g)	CHOLESTEROL (mg)
Goat	122	2.6	0.79	23	63.8
Beef	179	7.9	3	25	73.1
Pork	180	8.2	2.9	25	73.1
Chicken	162	6.3	1.7	25	76
Lamb	175	8.1	2.9	24	78.2

Source: US Department of Agriculture figures, 2001

and those figures are continuing to grow. Next came India, followed by Pakistan, Nigeria, and Sudan. These five countries accounted for around 62 percent of global goat meat consumption that year.

While not as widely eaten in the Western world, the demand for goat meat is rapidly rising, partly due to the needs of ethnic groups and partially because people are discovering its delicious flavor. Many countries unable to supply demand with domestically raised goat meat, import from the world's largest exporter: Australia. According to Meat & Livestock Australia, the value of goat meat exported in 2017 had jumped 38 percent from the previous year to A$277 million. The United States was and still is its biggest customer, and overseas sales have more than quadrupled in a decade. Nearly all of the meat exported from Australia comes from farmed or captured feral goats (see pages 26–27).

WHO EATS GOAT MEAT?

Cabrito, slow-roasted or barbequed milk-fed baby goat, is a specialty in Latin America and parts of the American Southwest. In Okinawa, thinly sliced goat meat is served raw as *yagisashi*. The Nepalese love goat meat and serve it in several offal-based dishes, and roast goat is staple fare at Easter celebrations in southern Italy, Portugal, and Greece; it's also served in the north of Portugal on Christmas Day. And chevon is popular for use in hearty fare like Jamaican curries and goat head soup.

According to IndexBox, the leading consumer of goat meat today is China. In 2015, consumers there raised or purchased 2,496,000 tons (2,264,000 tonnes) of goat meat

Right: *Farmers in Australia and New Zealand gather feral goats to fatten for sale as meat or, occasionally, to use to develop on-the-farm breeding programs.*

Goats for Fiber ✐

Although less familiar than the wool provided by sheep, goats have long been bred as a source of fiber. Fiber goats produce two luxury fibers, both with a strong worldwide market—mohair and cashmere.

MOHAIR

Mohair is the silky fiber produced by Angora goats (see page 188); the name comes from the Arabic word *mukhayyar*, which once described a high-quality cloth made from goat fiber. Angora goats originated around 2,000 years ago in Turkey. Although they had been imported into Europe as early as the sixteenth century, they failed to thrive. However, a shipment of Angoras to South Africa in 1838 fared much better. Thousands more joined them as the century passed, building what is today the world's leading mohair industry. The second-largest producer is the United States, where the industry can trace its origins back to the gift from Turkey of seven Angora does

Below left: *A shearer's helper on the Edwards Plateau, Texas, gathers up mohair, 1940. Historically, ranchers here maintained Angora herds of 1,000 or more. They reported a population of 2 million in 1992 alone.*

Below: *Angoras have spread around the world from their original home in Turkey. These goats are on a free-range farm in Africa.*

Mohair throw

Mohair yarn

GOATSKIN

Though goats aren't raised strictly for their skins, goatskins are a lucrative by-product in many parts of the world. Soft, pliable, strong Moroccan goatskin leather, for example, is widely used for gloves, wallets, shoes, and book binding. It was also once used for crafting parchment. Goat suede is another soft, supple leather used for clothing because it's strong, lightweight, and comfortable to wear. And entire tanned hides, sometimes with the hair left on, were once used to make waterproof canteens, like the *bota* bag of Spain (shown below) and those used in Eastern Europe to ferment kefir. (Goatskin is also used to make musical instruments, see pages 152–53.)

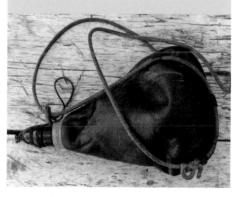

and two bucks in 1849. Angora goats were also exported to Australia and New Zealand in the 1800s, leading to strong industries in both of those countries, too.

Today's Angora goats have been selectively bred for prolific fiber production—even their faces and legs are covered with fleece. A typical goat grows between 6 and 30 lb. (2.7–13.6 kg) of fiber a year. Because the fiber grows all year round, goats are shorn twice a year, when their locks reach 4 to 6 in. (10–15.2 cm)—usually 2 to 4 weeks prior to kidding, then again six months later.

Mohair is composed of mostly keratin, the protein that is found in human hair and nails (and goat hooves and horns). It grows as wavy ringlets on the goat, resembling very fine human hair, and it is non-flammable, stretchy, durable, and warm. The diameter of a fiber ranges from 25 to 45 microns (one millionth of a meter); young does' fiber should measure no more than 31 microns, and young bucks, less than 33. Diameter also increases as a goat ages, with the ultra-fine mohair from kid goats (measuring 27 microns or less) fetching the highest prices. The fiber can be used as is, or blended with other fibers, and employed to make clothing, home furnishings, and even fur for teddy bears.

CASHMERE

All goats grow two-layer winter coats—
the one that you see (guard hair), and a
downy undercoat that starts growing as
fall approaches, achieves full length by
midwinter, and is shed over several
weeks each spring. Some goats grow an
abundance of ultra-fine undercoat. This
is cashmere, a luxurious fiber that has
been used for centuries to craft strong,
lightweight, wrinkle-resistant garments.

The world's first cashmere industry
was founded in Kashmir, and its primary
product was the cashmere or "pashmina"
shawl, a garment that took the Western
world by storm in the eighteenth century.
Napoleon famously presented a pashmina
shawl to his Empress Josephine, and the
empress so fancied the garment that her
wardrobe eventually included 1,000
shawls. Queen Victoria of England was
another famous collector. In fact, so great
was her enthusiasm for cashmere that the
Shah of Persia presented her with a herd
of Cashmere goats upon her ascension to
the throne in 1837 (see page 150).

Today, China is the largest producer of
commercial cashmere; Mongolia, Nepal,
India, Iran, Afghanistan, Australia, and
New Zealand are big producers, too.

Cashmere goats are a type selected for
prime cashmere production, rather than
a specific breed. They can be long- or
short-haired since the length of the guard
hair has little bearing on the length or
quality of the cashmere undercoat. All
colors are acceptable but the shearable
parts of the goat (excluding face, lower
legs, and belly) should be the same
color. The size of the goat as well as the
density, length, and fineness of its fleece
determine quantities, but the average goat
yields 4 to 6 oz. (113–70 g) of cashmere
a year.

Like mohair, cashmere fiber is
measured in microns, the industry
standard being fiber that is less than
19 microns in diameter. And fiber length
must be a minimum of $1\frac{1}{4}$ in. (3 cm).
Cashmere is low luster, ultra-soft, and
crimped rather than wavy, a naturally
insulative trait that gives it loft. This
results in garments that are soft and
lightweight but warm in the winter
and cool in the spring.

Left: *An 1805 portrait
of Empress Joséphine
by the artist Pierre-Paul
Prud'hon. First wife
of Napoleon Bonaparte,
Joséphine was a
preeminent fashion setter
who so loved pashmina
that she paid 20,000
francs for a single shawl.*

Below: *Ultra-fine
pashmina cashmere is
still produced in and
exported from Kashmir.
These Kashmiri goats
and their goatherd are
roaming through the
mountains of Ladakh
in northern Kashmir.*

Above: *In more recent years, two new breeds— the Nigora and the Pygora (shown here)—have been created for fiber. For more details on the fleece of each breed, see pages 191 and 200, respectively.*

Below: *Goat hair has been used for hundreds of years to make tents that withstand challenging conditions. These traditional Bedouin tents are in a camp in Wadi Rum, Jordan.*

HOW MANY GOATS DOES IT TAKE?

According to the international Cashmere and Camel Hair Manufacturers Institute, which represents the interests of producers and manufacturers, it takes the annual production from at least two goats to make a two-ply cashmere sweater, and the cashmere from four to six goats to craft a sports jacket.

HAIR

Goat hair is a coarser product than either cashmere or mohair, and is usually a by-product from goats raised for slaughter. Hair is generally too coarse to be woven into modern-day textiles—and according to Roman author Pliny the Elder, only the poorest people wore goat-hair cloth—but it has had numerous uses throughout history. It was used for making hair shirts, for example, worn next to the skin to show repentance. The Romans also used it for sacking, sails, and ropes.

Black goat hair was used by nomadic tentmakers in the Middle East, North Africa, and Tibet—black because it provides greater coolness during the heat of the day, and goat hair because when its fibers get wet they expand, forming a waterproof enclosure. Turkish nomads raise black Kil-keçi hair goats to this day.

There is still a market for goat hair today. It's imported to Europe from China, Mongolia, and the Indian sub-continent and used to fashion brushes, cloth for suit interlinings, and as a binder when mixing plaster.

Goats for Brush Control ~

Goats are natural climbers, unfazed by steep slopes or ditches, and they are able to eat their way through dense brush. These qualities make them ideal when it comes to clearing brush in hard-to-reach places, or in areas where toxic chemicals and herbicides are undesirable.

Landowners sometimes informally turn goats into fenced areas to address hard-to-rout vegetation while providing a portion of the animals' nutritional needs. Or they pasture them with different species like horses or cattle to consume brush and weeds those species can't eat. But the brush-clearing goats making the news today are the battalions of goats provided by rental companies to clear terrain for a fee.

Left: *Excess goats have always been used for brush control, hence the semi-derogatory appellation "brush goat" for a goat of nondescript looks and heritage.*

RENT A RUMINANT

Goats are a cost-effective and ecologically friendly way for cities, parks, natural resources commissions, and private-property owners to create fire breaks and to clear unwanted vegetation. Typically a company evaluates the property to ascertain what sort of vegetation is involved, how much of it there is, how many goats the property can hold, and the logistics of setting up goats in that location. Some charge a flat fee per acre cleared; others charge a set amount per goat, per day.

The company supplies fences (electrical, usually, to keep goats in and predators out), temporary shelters, tanks for water, any supplementary feed needed, and sometimes livestock guardians (see page 109).

The type of goats used varies, but they must be healthy, good travelers, docile, and easy to catch. Dairy–meat crossbred wethers are prized for their size and hardiness, whereas kids and pregnant or lactating does require more maintenance. A general rule of thumb is that ten goats can clear an acre in one month. Some crews are larger—numbering 100 goats or more—but the more goats present and the longer they stay, the more likely they are to damage desirable vegetation like trees.

Clearing with goats is not practical in situations where lots of desirable vegetation is mixed with vegetation that is undesirable (see box below), or when a good proportion of the vegetation is toxic or poisonous to goats. The service isn't a one-time thing, either—a goat cleaning crew may have to return to the same area several years running to fully destroy hardy brush and weeds. But in the process their droppings provide nutrient-rich organic matter that boosts soil fertility, texture, water-holding capacity, and infiltration.

Below: *Goats can't safely eat all plants, though they savor some that you wouldn't think they could eat, like brambles, wild roses, and poison ivy.*

SAFE VERSUS UNSAFE

Goats can safely eat:	Keep goats away from:
Dogwood	Rhododendron
Sumac	Yew
Mesquite	Oleander
Willow saplings	Laurel
Kudzu	Laburnum
Poison oak and poison ivy	Jimsonweed (datura)
Phragmites	Water or poison hemlock
Knotweed	Deadly nightshade
Wild roses	Locoweed
Ragweed	Lupines
Thorny brambles, including wild blackberry and raspberry	Daffodils
	Azaleas
	Foxgloves

Packgoats

Goats are sure-footed, agile climbers, and they can go anywhere a human leads them—even steep, rocky terrain, unlike bulkier pack stock like equines and llamas. They have been used to carry loads from place to place since early domestication, and they still do in parts of the world today, but nowadays they're commonly used for recreational packing.

REASONS TO CHOOSE PACKGOATS

Camping or trekking with a packgoat in tow, carrying equipment in panniers attached to a saddle, offers many advantages, and as a result is gaining popularity both in North America and in parts of Europe.

Goats will fall in line behind a leader, and they naturally bond with humans, staying with them of their own volition, so they needn't be kept on a lead or picketed at night. They're also clean, biddable, easy to handle, and relatively close to the ground—even children are able to hoist packs onto their backs.

Goats are an eco-friendly choice, too, simply nibbling a leaf here, sampling a twig there, and never stripping a single plant of its foliage. They browse most or all of their meals, and then process what they eat into dry, deer-like pellets that are deposited on the go, instead of in unsightly piles.

Packgoats are inexpensive to buy, outfit, transport, and maintain. They're easily transported in vans, SUVs, and "goat totes" in the back of pickup trucks; no expensive trailer is needed. They're strong, too. A mature, well-conditioned goat can carry a quarter to a third of its own body weight up to 15 miles (24 km) in one day—the average packgoat weighing 200 lb. (90 kg).

And lastly, friendly goats are wonderful camping companions. Their entertainment value alone is priceless.

Below: *Goats have been used around the world to carry loads on their backs since the earliest times. This caravan of goats is trekking through the challenging terrain of western Nepal.*

SELECTION AND TRAINING

Any willing goat can carry a pack. Even miniatures like Nigerian Dwarf and Pygmy goats are fine for day trips, but dairy goats are the packgoat of choice. Their longer legs and greater agility make them more suited to steep, rocky climbing than shorter-legged, broader meat goats—and some packers choose milking dairy does that can provide milk along the way, although these need protection for their udders. Meat goats, however, can carry more weight and they do well in less rugged terrain. Whatever the breed, most goat packers prefer wethers but a dry (non-lactating) doe works just as well. And packgoats generally have horns because these act as thermoregulatory organs and help cool the animals in hot weather.

Starting with bottle-fed kids makes training the essence of simplicity. Kids as young as 3 and 4 weeks old follow humans and learn to walk through puddles or climb over rocks and logs. Some goat packers fit their goats with lightly loaded, soft training packs as early as 6 months of age, although there are those who claim that this can damage a young goat's spine. Most packers will take their goats camping, but will not add packs until the goats are almost 2 years old.

Above and below:
Packgoats are intrepid travelers and the perfect solution for anyone who loves the outdoors but can't handle its physical challenges. Goat packing gear doesn't cost a fortune, and nor does a good goat to carry it.

Harness Goats

Throughout antiquity goats were used in harness to convey people and goods from place to place. In the nineteenth and early twentieth centuries, goats were used for pulling children's and invalids' carriages, and in any farm situation, a wether or buck could earn his keep by pulling light machinery and still provide tasty meat at the end of his working life.

GOATS FOR HIRE

Goat cart concessions were once a fixture in city parks from New York to Paris to Golden Gate Park in San Francisco, as well as at British seasides, including the one at Brighton Beach where, in the mid-1800s, rent for a goat and wagon cost a shilling an hour.

Below: *Coney Island in New York was just one of many seaside attractions in the United States where tourists could enjoy a goat-drawn cart or carriage ride for pleasure.*

GOAT MEN

During the early and mid-twentieth century, a few intrepid Americans used goats to travel long distances, the most famous of these being the "goat men," Overland Jack and Ches McCartney.

When Virginian John Rose, later known as Overland Jack, was nine years old, he fell beneath a freight train and lost both of his legs. In 1909, at age 20, he took to the road driving a small wagon drawn by four Spanish goats, earning his way by carrying out minor repairs, sharpening tools, and selling picture postcards of himself and his goats. The cards told the story of his life on the road. Overland Jack died in 1963 and was buried in Big Sandy, Texas, where a goat head carving and a picture of Jack with his goats and wagon adorn his tombstone.

Charles "Ches" McCartney was born in 1901 in Iowa, where, he later claimed, his only childhood friends were the family goats. As an adult he also took to the road—as an itinerant preacher in a rickety wagon pulled by a team of goats. Bucks and wethers pulled the wagon, while kids and does rode inside, and Ches supplemented his diet with their milk. Between 1930 and 1968 Ches and his goats visited all of the continental United States. During his years on the road, Ches posed for pictures and sold a wide assortment of picture postcards that remain popular collectables today.

Above: *Ches McCartney with his team of goats and a wagon of wares, photographed on the road in the 1920s.*

VINTAGE PHOTOS

Another type of desirable collectable is the photograph of children posing with a goat and cart. In 1919, Chicago Ferrotype Company's Mandel-ette postcard-format camera came on the market, capable of creating picture postcards on demand. Itinerant children's photographers snapped them up and took to the road.

Some of these photographers took a saddled pony or donkey with them but far more popular were the goat cart men. Until the early 1960s, untold numbers of beaming youngsters posed for their portraits sitting in a goat cart or wagon, holding the reins.

Goats were chosen for their placid temperaments and good looks. Most had impressive horns, and horned Angoras were especially popular. Photographers usually worked a specific area, so collectors often spot the same goat in numerous photos. Carts and wagons also often featured a plaque or lettering, signifying the year a photo was taken and in what town.

Novice collectors sometimes find it hard to determine whether photos are the work of a traveling photographer, or a depiction of children genuinely working with goats. The giveaway is usually the child's hands—if the reins are hanging loosely and the child is focused on the photographer instead of the goat, the picture was probably staged.

Above: *A turn-of-the-century photographer captures a couple on honeymoon on the beach at St. Augustine, Florida, alongside a posed harness goat.*

Right: *Goat carriages offer pleasure rides in New York's Central Park (middle), and three children perch on a goat-pulled cart in upstate Rochester.*

Right: *Harry and David of Team Snazzy Goat—two American Cashmere harness goats, chosen for their trim, muscular build.*

DRIVING GOATS TODAY

Goats are still used in some parts of the world to work on farms. In addition, goat driving for pleasure is becoming an increasingly popular pastime. A healthy adult goat can pull up to one and a half times his own weight, depending on the harness and vehicle used, and the job at hand. Large, mature dairy-breed wethers are said to make the best driving goats, but with the right equipment and training, any goat—even miniature goats—can pull size-appropriate carriages or wagons. Goat harnesses made of leather or nylon, and carts and wagons designed for goats, are readily available, although some goat

Below: *Although several types of harness are used, most harness goat aficionados prefer a basic light driving harness like this one, especially for pulling carts, carriages, and small wagons.*

owners choose to craft their own using the vast array of information available online.

Goats can pull singly or in tandem. Most commercial carts are designed to be pulled by a single goat, and singles can also pull small wagons. For heavier work, teams are better.

Most beginners start with a two-wheel cart and a pleasure harness. A goat cart is easy to use and easy to care for, but it can't manage sharp turns without tipping and it must be perfectly balanced to save the goat's back. To check balance, one person sits on the seat while another person lifts the empty shafts. When the shafts are level, they should almost seem to float, exerting little or no pressure on the goat's back. A better choice in many cases is a four-wheeled wagon or lightweight carriage. These are considerably more stable and easier on the goat than a poorly balanced cart.

A BASIC GOAT HARNESS

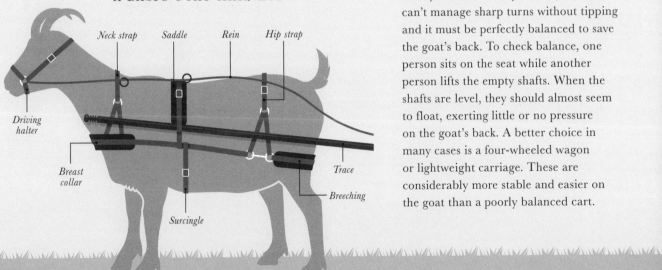

Neck strap · Saddle · Rein · Hip strap · Driving halter · Breast collar · Surcingle · Trace · Breeching

Goats in Mythology & Folklore

Goats were once such an integral part of life that the ancients incorporated them widely into their beliefs.

GOATS AMONG THE GODS

Albasty, or Alvardy, the Azerbaijanian goddess of birth and prosperity, was associated with goats and could in fact change herself into a goat. Even today in Azerbaijan, a soft, protective rope woven of goat hair and fleece is stretched around the head of the bed in which a woman gives birth.

And from a different tradition, Agni, the ancient Hindu god of fire, was often pictured with flaming hair, black skin, with two heads and four arms, and seated

Below: Some say that Agni, the Hindu Vedic god of fire, rides a black ram, while others say he rides a black buck goat like the one shown here.

NOTABLE GOAT DEITIES

Akerbeltz (Basque deity): from *aker* for "he-goat," and *beltz* for "black"; protector of goats and other animals. Basque farmers sometimes kept a black buck in their stables to honor Akerbeltz and protect their other livestock.

Chernobog (Western Slavic god of the dead): said to ride a goat, and to take the form of a black he-goat on occasion. His followers set themselves apart by wearing goat tattoos.

Dažbog (Slavic sun god): sometimes perceived as a white goat.

Marduk (patron deity of Babylon and Sumerian god of magic): associated with goats. Goats were considered uncanny beings due to his patronage.

Mari (Basque goddess): often took the form of a goat and was said to be partial to black he-goats.

Yang Ching (Chinese goat god): depicted with a goat's head and dressed in goatskin; he protected his followers from wild animals.

Above: *Thor, the hammer-wielding Norse and Old Germanic god of thunder, rumbles across the sky in a chariot drawn by two goats, Tanngnjóstr and Tanngrisnir.*

on a black goat. Goats were his preferred sacrificial animal.

Several deities also rode goats or hitched goats to their chariots, the best known being Thor, the Norse god of war and thunder, whose chariot was drawn by a pair of goats—Tanngnjóstr (Gap Tooth) and Tanngrisnir (Tooth Grinder). Thor could slaughter his goats for his evening repast, then gather their hides and bones and set them aside, knowing that they'd regenerate again by morning's light.

Goats were also sacred to the Greek goddesses Artemis (goddess of the hunt) and Juno (goddess of marriage), as well as to their Roman counterparts, Diana and Hera, respectively.

GOAT-MEN AND GOAT-WOMEN

A host of ancient deities, fairies, and spirits were said to be part man or woman, and part goat. Classical Greek and Roman mythology, for example, is rife with goat–man hybrids. Consider Pan, the Greek god of woodlands, shepherds, goatherds, and rustic music, and also his Roman counterpart Faunus. Both had the hindquarters, legs, beards, and horns of a goat, and live goats were often sacrificed to them. Lecherous Greek satyrs had the ears, horns, beards, and tails of goats, and in the theater they were represented by actors wearing goatskins.

Greek panes, called fauns by the Romans, were sometimes said to be sons of Pan or Faunus. They were the spirits of wild places and, like their fathers, they were part goat and part man. One of these was Aegipan, who, according to one story, aided Zeus in battle and was rewarded by being placed in the heavens as the constellation Capricorn.

The folklore of the British Isles also includes numerous creatues assuming whole or part goat form.

Puck, who is also known as Robin Goodfellow, was a mischevious goat-footed sprite, or hobgoblin, who sometimes helped people and sometimes played nasty tricks. He's best known for his appearance in Shakespeare's *A Midsummer Night's Dream.*

In Scotland, the glaistig was a female water fairy with the legs of a goat. As long as she wasn't provoked, she was happy to tend a farmer's cattle and watch over children and old folk. Scotland was

Left: *Pan playing his customary pipe, in an 1862 engraving by the British artist Frederic Leighton. Pan became a popular figure for Romantic artists of the eighteenth and nineteenth centuries.*

AMALTHEA

Amalthea was the mythical she-goat who suckled the infant Zeus (shown right) in a cave on Crete's Mount Aigaion ("Goat Mountain"). While still a baby, Zeus accidentally broke off one of Amalthea's horns, which would later become the cornucopia, or horn of plenty. When Amalthea died, Zeus saved her skin—which became the aegis (a protective garment worn by the goddess Athena)—and placed her among the stars as Capella (little goat), part of the constellation Auriga.

also home to the urisks, or brownies, who also enjoyed minding cattle, as well as helping humans with their daily chores. The urisks are often depicted with the upper body of a human and the lower half of a goat.

And in Wales, the fairies known as gwyllion also assumed goat form. They were said to visit stables and comb the goats' beards to make them presentable for special occasions.

THE YULE GOAT

The Yule goat has been part of midwinter lore in Scandinavia and Northern Europe for more than 1,000 years. Initially associated with Thor (see page 143), the harvest, or fertility rites, early Christian missionaries frowned on pagan revelry and demoted the Yule goat to demon status. Eventually, however, it evolved into a benevolent being, often depicted pulling the sleds of Christmas fairies.

Goat ornaments made of straw bound with red yarn (called a *julbock* in Sweden) are now popular at Christmas time, and in many places huge straw *julbocken* are erected to usher in the holiday season. The largest and oldest of these is the *julbock* erected each year in the city of Gävle, Sweden, typically constructed with around 4,000 ft. (1,200 m) of Swedish pine.

Above: *The Gävle Goat is a tradition dating back to the 1960s. This giant* julbock *is erected on the first Sunday of Advent in Gävle's Castle Square.*

Left: *Early twentieth-century Scandinavian artists like Jenny Nyström popularized the Yule goat and his* tomten *handlers through hundreds of charming, antique Christmas cards that are treasured collectables today.*

The Sacrificial Goat

Offerings to the gods have been an important aspect of religious practice since ancient times, intended to appease, to ask for favors, or to offer thanks for favors rendered. In many instances, the sacrificial animal has been a goat.

THE ANCIENT GREEKS AND ROMANS

Followers of Dionysus, Apollo, and Aphrodite sacrificed goats to their deities, whereas, because goats were sacred to Artemis (Diana) and Juno (Hera) (see page 143), their followers generally did not. Goats were bathed, bedecked with ribbons, and taken in a procession to the temple. At the altar, water was poured on the goat's head to make it nod and thus agree to the sacrifice, before a priest deftly slit its throat. Sometimes huge numbers were sacrificed. Before the Battle of Marathon in 490 BCE, Athenians promised to sacrifice goats equal to the number of Persians killed in the battle, offering up 500 goats each year until the total was met—a ritual that was still being carried out 90 years later.

HINDUISM

In the Vedic Age (c. 1500–500 BCE), goats were commonly sacrificed to all Hindu gods and goddesses. The Bhagavad Gita, however, dating to between the fifth and second century BCE, forbids animal sacrifice and it's rarely practiced today, except in Nepal and some of the Eastern states of India.

Above and below:
The festival of Lupercalia involved the sacrifice of a dog and a goat in Rome's Lupercal, the cave where Romulus and Remus were believed to have been raised by a she-wolf.

Bali (animal sacrifice) is a sacred rite still practiced by certain followers of Shakti—"the great mother" who takes many forms, including the goddess Durga. Goats are still sacrificed in great numbers at Dashain, a 15-day Nepali feast commemorating the victory of Durga over a water buffalo demon. Goats have a vermillion dot (*tika*) painted on their forehead before decapitation with a single blow from a large machete or an ax.

JUDAISM AND CHRISTIANITY

The Hebrew Torah and the Old Testament of the Christian Bible tell us that goats were popular sacrificial animals in those religions. Instructions for performing sacrifices are found in the Book of Leviticus, and detail the means of burning either parts or the whole of the animal. And the Torah describes the offering of a sacrificial animal (*qorban*), which had to be kosher (a goat, sheep, bull, dove, and so on but no unclean animals like swine) and had to undergo Hebrew ritual slaughter, severing the trachea, esophagus, and large blood vessels in the neck with one swift pass using a special knife.

ISLAM

Qurbani is the sacrifice of a livestock animal during Eid al-Adha, a holiday marking the end of the Hajj pilgrimage and honoring the willingness of Abraham to sacrifice his son Ishmael, in obedience to Allah's will. Many animals are sacrificed, including sheep, cattle, and camels, but goats are such a common sacrifice that the festival is sometimes called Bakra-Eid ("Goat Eid") in Urdu.

Each animal is slaughtered separately with an ultra-sharp knife, out of the sight and sound of other animals. The meat is then divided into three equal parts: one for the owner's family, one for relatives and friends, and one for the poor.

SANTERÍA

Animal sacrifice, *eyebale*, is an important tenet of Santería, marking life events such as births, marriages, and deaths. While poultry are the most commonly sacrificed animals, larger species including goats are used for major events. They are killed by cutting the carotid arteries with the single stroke of a sharp knife, then eaten as an act of communion with the spirits.

Above right: Goats are a favorite animal to offer to Allah at Eid al-Adha. Goats slaughtered at Eid are usually raised by the families that slaughter them, so the act is a true sacrifice.

Military Mascots ✑

Animals have always served as mascots in the military, and goats are no exception—they traditionally served aboard naval battle ships until recent times, but they have served with land-based units, too, and some individuals have gone on to achieve legendary status.

SERGEANT BILL

At the start of World War I, a unit of new recruits to the Canadian Expeditionary Force traveling through Saskatchewan spotted a young Daisy Curwain and her goat, Bill. They asked her if they could have Bill for their goodluck mascot, and she agreed.

Bill traveled with the unit to England aboard the SS *Lapland*, but the unit was then ordered to France—without Bill. As Sergeant Harold Baldwin later recalled,

Above: *British soldiers at the Western Front during World War I, accompanied by their pet goat.*

Left: *"Pitch" the goat mascot befriends the son of Admiral William T. Sampson, aboard the USS New York, in around 1899.*

"We could not part with Billy; the boys argued that we could easily get another colonel, but it was too far to the Rocky Mountains to get another goat." So Bill was smuggled to the front, where he distinguished himself in many battles. After the second battle of Ypres, Bill was found in a crater shell guarding a Prussian soldier, despite being wounded in the neck. He then went on to suffer from trench foot and shell shock, and was wounded at Festubert, where he head-butted three soldiers into a trench just before a shell exploded precisely where they'd been standing.

Bill spent most of his four and one-half years of service in France and Belgium, and was promoted to the rank of Sergeant. He was also awarded the 1914/1915 Star, the British War Medal, and the Victory Medal before eventually being returned to Daisy Curwain. Today his body holds a place of honor in the Broadview Museum.

WILLIAM DE GOAT

Perhaps the most famous goat to serve in World War II was Air Commodore William de Goat, mascot of the British 609 West Riding Squadron. In 1944, William was among the first of his unit to land in France. The squadron was then deployed to the Netherlands, then Germany. William's adventures were many—being kidnapped, devouring important documents, and nearly killing himself by sampling the toxic paint used to decorate his horns.

For his time in service William earned Distinguished Service Order and Flying Cross medals and he continued gaining rank throughout the war. William celebrated his promotion to Wing Commander by ingesting 30 cigarettes, two bowls of chrysanthemums, and the Commanding Officer's mess bill. He eventually became the highest-ranking officer in the squadron, whereupon the unit's CO began saluting William before taking off on a mission.

Above: *A Royal Air Force pilot is shown attempting to take a reluctant "Wing Commander Billy" for a walk at the fighter station in Manston, England, in 1943.*

GOATS AND THE ROYAL WELSH

The British Army has a long tradition of maintaining mascots to raise the morale of soldiers at war and for ceremonial purposes during peacetime. Official mascots are considered British soldiers and are supported by public funds; they have a regimental number and rank and, depending on their behavior, can be promoted and demoted. Goats have always been a popular choice—with perhaps the most famous being the Welsh regimental goats.

As early as the late eighteenth century, the Royal Welsh Fusiliers were presenting themselves for review with a decorated horned goat leading the way. Then in 1844, Queen Victoria gifted the Fusiliers with a royal goat from her own herd of Cashmeres at Windsor Park (see page 132) to use as their mascot, followed by a second gift in 1862, to the Welsh Regiment of Infantry. Finally, in 1881,

on becoming the 2nd Battalion, the Welsh Regiment took up the same custom. This began a tradition of mascots (drawn from the Royal Herd, whenever possible), with the same name passed from one goat to its successor in each case.

Over recent years, battalion restructuring has led to the formation of the Royal Welsh, with the continuation of two goat mascots—Llywelyn with the 1st Battalion, and Shenkin with the 3rd Battalion. Whenever a mascot dies, notice is sent to the Queen, together with permission to select a replacement—most recently from the herd at Great Orme at Llandudno (see page 27), which is known to have Windsor blood in its ancestry. Most mascots have been bucks, although Shenkin IV—the current Shenkin— is a wether.

Left: *White Cashmere goats have long been favorite mascots for various Welsh regiments. This was Billy of the Royal Welsh Fusiliers, photographed on St. David's Day in 1951.*

Below: *Shenkin II of the Royal Welsh 3rd Battalion, with his handler, Goat Major Mark Jackson.*

GOATS AND BEER

Bock is a dark, robust German lager dating back to the fourteenth century. Originally brewed in Einbeck in the north, it gradually moved south to Munich, where the dialect of Bavarian brewers resulted in the pronunciation "Ein Bock," bock meaning "buck." Thus its marriage to goats was established, and since the mid-nineteenth century, bock beer labels have featured goats. But the connection does not just relate to labels. Following in the British tradition of feeding horses with Guinness (a stout that supposedly aids digestion, acts as an antioxidant, and offers a supply of B-vitamins, iron, and other minerals), goat owners often give their animals dark beer to boost their energy. Some give it to does to give them a lift after kidding. The usual dose: one full bottle, repeated once if needed. It was traditionally dosed using a long-neck glass bottle. Nowadays most let the beer go flat and administer it with a dose syringe.

Mascots join the battalion at the rank of fusilier and have their own full-time trainer and handler (the Goat Major), but a supposed daily diet of two cigarettes (to eat) and a draft of Guinness (see box) is a tradition that has fallen by the wayside!

Although Welsh regimental goats accompanied their battalions on active service in the past and were, like Sergeant Bill and William de Goat, awarded the appropriate service and campaign medals, the last to see active service accompanied his battalion to Korea in 1951. Today, these goat mascots are responsible simply for leading military parades and acting as ambassadors.

Right: *A Welsh Army regiment on parade in 1918, with a white Cashmere goat at its head. This ceremonial role continues with the Royal Welsh today.*

Goats & Musical Instruments

Nowadays musical instruments are generally fashioned using synthetic materials, but in days past they were often made using the skin or horns of goats.

BAGPIPES

When people think of bagpipes, they rarely think of goats, yet most bagpipes were originally crafted with air bags made of goatskin—sometimes with the hair off, sometimes with hair on, and sometimes, as with the Greek *tsampouna*, with the hairy side turned inside to help regulate moisture within the bag.

Pipers in the Lubusz region of Poland play large, low-pitched bagpipes called *koziol bialy* (white goat) and *koziol czarny* (black goat). The bag of the *koziol bialy* is formed from a white goat's hide, with the fleece facing outward; the bag of the smaller *koziol czarny* is made of hairless black goatskin. Both usually feature carved wooden goat heads on the stocks of their chanters.

Another type of traditional bagpipe is the *bock* of Central Europe, now mostly made in the Chodsko region of the Czech Republic. Its bag is also made of goatskin, and its chanter stock is usually accented with a carved wooden goat head, complete with eyes, ears, and horns.

Below left: *An eighteenth-century engraving of a Polish bagpipe player. The bags of ethnic bagpipes are usually crafted using goat or sheepskin, sometimes with the hairy side of the pelt inside the bag.*

Below: *Whole goatskins hang in readiness to be made into bagpipes. Goatskin is said to be thin yet strong, so well suited to forming the bag of the instrument.*

THE GOAT BASS

The goat bass, or *kozobas*, is a bowed instrument dating back to the 1960s, used to play folk music in Ukraine. Although not made from real goat, it consists of a fret board attached to a carved wodden goat peg head, with a drum at the other end. The electrified goat bass played by Zahyney Volodymyr, of the popular Ukrainian "Ukrabilly" band Ot Vinta, is a particularly sophisticated example.

DRUMS

Drums in many parts of the world have traditionally been crafted using goatskin to form the drumhead, with the drum itself crafted in a range of different shapes. Notable frame drums (circular, and wider than they are high) include the *bodhrán* and *tabor* of Ireland and Wales. Goblet drums (upstanding drums with goblet-shaped bodies) include the *darbuka* of Turkey, the *djembe* of West Africa, and the *doumbek* of North Africa, Egypt, and the Middle East. And the double-headed *dhol* is a drum type found throughout India and Pakistan.

THE BUKKEHORN

Goat horns have been used for crafting signaling horns for millennia, among them an ancient Scandinavian musical instrument called the *bukkehorn* ("buck's horn"). In Norway, finger holes were added to the basic trumpet design as early as the Northern Bronze Age (1700–500 BCE) so that music could be played on the horn. And in around 1700 CE, a version with a reed was added to the group, to be played like a clarinet. Bukkehorns were traditionally used by shepherds and goatherds as signaling or scaring instruments that made music on the side.

Above: *The bukkehorn is only one of many horns crafted using the horn of a goat. Even shofars are sometimes made of goat horn, especially in Holland.*

Below left: *Goatskin makes a perfect drumhead, as for this Turkish darbuka, but it's used in making other instruments, too. A case in point: banjo heads are usually crafted of thin, tough goatskin.*

FOLK SONGS

There are countless folk songs featuring our friend the goat. One of the best known is "Pat McGinty's Goat," a tune of 20-plus verses about a troublesome Irish goat, written in 1917 by British songwriters Bert Lee and R. P. Weston. It was performed in music halls by a contemporary American singing duo the Two Bobs, and later recorded in 1964 by Irish singer Val Doonican. It went on to become a popular American camp song. It begins:

"Now Patrick McGinty, an Irishman of note,

Fell in for a fortune and he bought himself a goat.

Says he, 'Sure of goat's milk, I'm going to have me fill.'

But when he brought the nanny home, he found it was a Bill."

Goats in Popular Culture

Goats have featured in pop culture—in books, cartoons, movies, and more—for a long time, but they've never been more popular than they are today, judging from their increasingly frequent appearances in social media, and the high-profile individuals with their own huge followings on YouTube, Instagram, and Facebook.

Left: *Over the last couple of decades, goats have appeared regularly in both movies and television programs.*

Below: *Goats have featured as characters in children's storybooks for even longer (see page 157 for a selection of some of the most popular titles).*

Far left: *Sparky, the goat star of the 2017–18 Broadway revival of the musical* Once on this Island.

Left: *The increasing presence of goats in all forms of social media attests to their growing popularity.*

FAMOUS REAL-LIFE GOATS

Happi: a Nigerian Dwarf goat who on March 4, 2012, set the Guinness World Record for the farthest distance traveled by a goat on a skateboard (118ft./36 m). She would have gone farther had she not hit a parking barrier.

Lucy: an American Pygmy goat that gave birth to six kids on March 12, 2006. It was, according to the Guinness Book of World Records, the largest litter of baby goats on record.

Murphy (or Sinovia): Because William Sianis (owner of the nearby Billy Goat Tavern) and his goat Murphy were ejected from a game of the 1945 World Series, Sianis was said to have placed the "Curse of the Billy Goat" on the Chicago Cubs, supposedly preventing the team from winning a World Series championship. (The curse was broken in 2016.)

Nanny and Nanko: Tad Lincoln's goats, who lived at (and sometimes in) the White House while Ted's father, Abraham Lincoln, was President.

Nirmala: The goat who traveled with Mahatma Gandhi, supplying him with fresh milk every day (below).

Mastana: a huge black-and-white Beetal goat (weighing 661 lb./300 kg), nominated the world's heaviest goat at the 2018 Faisalabad Goat Show in Pakistan.

Whiskers: Russell Harrison's pet goat (shown above), which lived at the White House when Russell's father, Benjamin Harrison, was President.

SPORT MASCOTS

Goats infrequently serve as mascots for athletic teams but there are some notable exceptions, one of them being Hennes, the mascot of the 1. FC Köln soccer team in Cologne, Germany (pictured right). The first Hennes (named after coach Hennes Weisweiller) was presented to the team in 1950 and immediately became a hit with fans; his likeness was integrated into the club's logo that year, and also provided the inspiration for the team's nickname, Die Geissböcke (the Billy Goats). When one mascot dies it is replaced by another, chosen by popular vote. The current goat—Hennes VIII—resides at the Cologne Zoo in between games.

Another famous sports mascot is Bill the Goat, mascot of the US Naval Academy's football team since 1893. Since 1905, most Bills have been Angora goats. The most famous—also called Three-to-Nothing Jack Dalton—served the team from 1906 to 1912 and got his name because Navy's Jack Dalton kicked field goals for 3-0 victories over Army twice during his reign. A group of volunteer midshipmen known as Team Bill care for Bill and his understudies and transport him to nearby games. Currently there are two Bills, both Angoras.

GOATS ON SCREEN

Chompy the Goat—the white town mascot from the Nickelodeon cartoon series *The Fairly OddParents* (2001–17).

Djali—Esmerelda's feisty goat in Disney's *The Hunchback of Notre Dame* (1996; shown below).

Franny—the wisecracking white Saanen doe in *Racing Stripes* (2005), voiced by Whoopi Goldberg.

Japeth—an elderly railway goat with detachable horns in *Hoodwinked!* (2006) and *Hoodwinked Too!* (2011).

Jeb—a grumpy, tin can-collecting goat in Disney's *Home on the Range* (2004).

Jimmy—the dynamite eater in "The Loaded Goat," an episode of *The Andy Griffith Show* (1963).

Mona—a sacred village goat in the South African comedy *Max & Mona* (2004).

GOATS IN CHILDREN'S BOOKS

Arashi no Yoru ni ("Stormy Night"; Yuichi Kimura & Hiroshi Abe, 1994): the first of a series of books, featuring Gabu the wolf and his goat friend, Mei.

Billy Whiskers: The Autobiography of a Goat (Frances Trego Montgomery, 1903): the first of more than a dozen books.

Klods-Hans (Clumsy Hans; Hans Christian Andersen, 1855): Clumsy Hans rides a billy goat to a palace in order to woo a princess.

Gandiv (Hariprasad Vyas, 1936–55): stories based on a good-natured caprine businessman Bakor Patel for this children's biweekly.

The Goat in the Rug (Charles L. Blood, Martin Link & Nancy Winslow Parker, 1976): features Geraldine, an Angora, whose fleece is used to weave a Navajo rug.

Greeley the Mean Goat (R. Barri Flowers, 2015): Greeley is found not be mean—just misunderstood and lonely.

A Hat Full of Sky (Terry Pratchett, 2004): features a doe called Black Meg and a buck called Stinky Sam.

Heidi (Johanna Spyri, 1880): features the goats Baerli, Schwanli, Türk, and Meckerli.

Huck Runs Amuck! (Sean Taylor & Peter H. Reynolds, 2011): stars a flower-loving goat.

The Wolf and the Seven Little Goats (Brothers Grimm, 1812): a mother goat leaves home to get food, warning her kids not to open the door to a wolf that wants to eat them (top right).

Three Billy Goats Gruff (Norwegian folktale): three billy goats pass across a bridge guarded by an ugly troll (right).

Zlateh the Goat and Other Stories (Isaac Bashevis Singer, 1966): an elderly she-goat saves a boy's life during a blizzard.

Goat Quotes & Proverbs

66 *That these animals are naturally friends to man, and that, even in uninhabited countries, they betray no savage disposition, is apparent from the following fact. In the year 1698, an English vessel having put into the island of Bonavista, two Negroes came aboard, and offered gratis to the captain as many goats as he pleased. The captain having expressed his astonishment at this offer, the Negroes replied that there were only twelve persons on the island; that the goats had multiplied so greatly as to become extremely troublesome; and that, instead of being caught with difficulty, they obstinately followed the men, like other domestic animals.* 99

GEORGES-LOUIS LECLERC, THE COMTE DE BUFFON,
Natural History, General and Particular (1781)

66 *There is, indeed, a fine independence of character about the goat which separates it by many parasangs [a unit of length] from the sheep. The latter lives placidly by faith, and seems assured of redemption; the former is possessed by the restless genius of unbelief. You cannot keep goats on the level road, even though you take the greatest possible pains to show them the farmyard and the fold at the end of it. They detest the commonplace. Rather than plod safely home by the regular way, they prefer to travel adventurously by paths of their own. By choice, they take the ups and downs of life; and when they do not find them ready made for them, they make them for themselves. If there is a heap of stones by the roadside, they get up on it—their spirits at once rise at finding themselves on an eminence.... If there is a ditch it is just the same. As many goats as possible get down into it and pretend it was in their way. They get among the sheep, and deliberately disorder the wooly procession. They jump over the sheep's backs to show their indifference to orthodoxy.* 99

PHIL ROBINSON, *The Poet's Beasts* (1885)

66 *If you catch hold of a goat's beard at the extremity, all the companion goats will stand stock still, staring at this particular goat in a kind of dumbfounderment.* 99

ARISTOTLE, *History of Animals* (4th century BCE)

PROVERBS

In a goat's eyes all things belong to goats.
CORNISH

A prudent man does not make the goat his gardener.
HUNGARIAN

The goat *must* browse where she is tied.
ROMANIAN

In time, even a goat can be taught to dance.
YIDDISH

A goat with long horns, even if he does not butt, will be accused of butting.
MALAY

Even the humblest goat is useful.
HAITIAN

Horns are not heavy for a goat.
ETHIOPIAN

An old goat will never learn to dance.
MOROCCAN

The more the buck goat stinks, the more the nanny goat loves him.
WALLOON

If you're short of trouble, take a goat.
FINNISH

The little billy goat has a beard, the big one has none. (Don't judge a book by its cover.)
JAMAICAN

If being well bearded brings happiness, a he-goat must be happiest of all.
ITALIAN

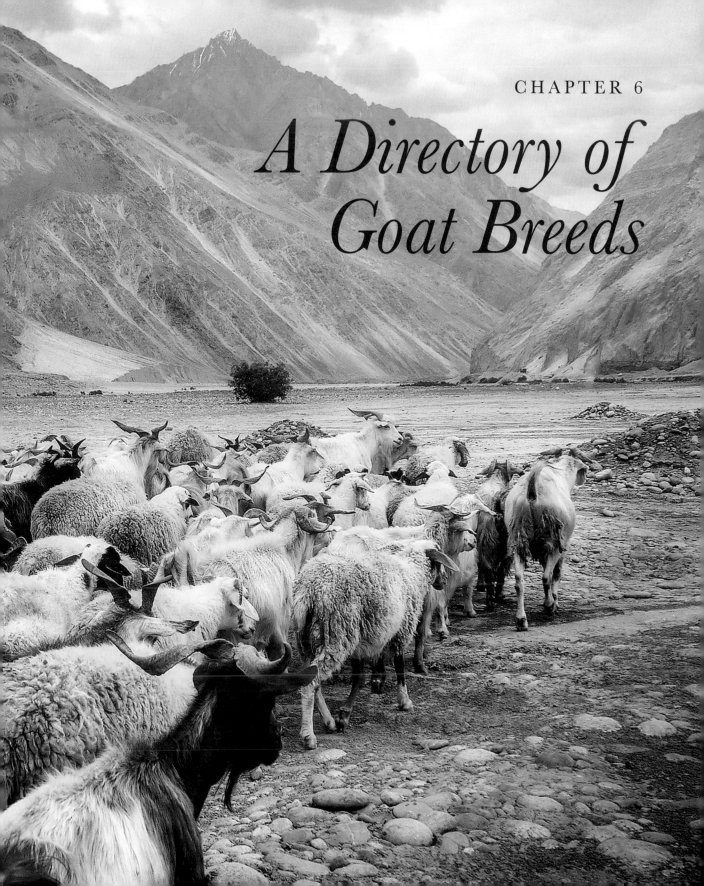

CHAPTER 6

A Directory of Goat Breeds

How Breeds Evolve ✒

A breed is a specific group of animals that look much alike, behave much the same, or have other characteristics that distinguish them from other animals of the same species. Breeds are formed by adaptation to their environment (landrace and feral breeds) or by selective breeding (improved breeds, or purebreds). Individuals of the same breed, when bred together, pass their distinguishing characteristics on to their offspring; this is a requisite for a breed.

Landrace is from German Landrasse, meaning "country breed." Landrace (sometimes called native) goats are domesticated, traditional types of goats developed over time by adapting to their environment (climate, disease, pests, and handling practices) in isolation from other breeds. They originate in a specific geographic area and, though genetically similar, they're physically less uniform than goats of a standardized or formal breed. They are not the product of selection for specific conformation, colors, size, or traits. Landrace goat breeds include, among others, Icelandic, Lappgetter, San Clemente Island, Spanish, and Old English and Old Irish goats. Some standardized, improved breeds descend from landraces and have "landrace" in their names—for example,

the Dutch Landrace, the Swedish Landrace, and the Finnish Landrace— but these are no longer true landraces.

Feral goats are descended from domestic stock that has returned to the wild and reproduced without human interference. When feral populations are isolated from the introduction of new animals and present for relatively long periods of time, some but not all authorities consider them feral breeds. Cases in point: the various types of

Below: The Dutch Landrace is one of several European breeds, including Icelandic, Swedish Landrace, and Lappgetter goats, that were developed with little or no influx from other breeds.

British feral goats—for instance, the Lynton goats of Devon in England, the Bilberry goats of Waterford in Ireland, and the New Orme herd of Cashmere (Kashmiri) goats in Llandudno, Wales.

Most purebred goats are members of standardized, improved breeds with herdbooks and breed standards: descriptions of precisely how outstanding individuals of the breed should look and perform. They are selectively bred for specific traits such as meatiness, milk production, or quality fiber. Usually only the best individuals are used to breed the next generation, causing loss of genetic diversity but creating a more uniform goat. Selective breeding includes monitoring health issues in a breed, like the G6-S hereditary defect in Nubian goats that causes delayed motor development, retarded growth, and early death. A quality selective-breeding program also avoids creating defects caused by breeding for "desirable" features, one example being extreme

HERITAGE BREEDS

A huge number of old-fashioned, lower-producing breeds of poultry and livestock all but disappeared during the 1950s when they were replaced by new and improved high-production breeds like the Holstein-Friesian cow and the various Swiss dairy goat breeds. This is why various heritage breed organizations were formed, notably the Livestock Conservancy in the US, and the Rare Breeds Survival Trust in the UK. Many heritage breeds are staging comebacks but it's a slow, ongoing process.

Roman noses in Boers, which can cause undershot jaws and lead to serious dental malocclusion. All types of breed mentioned above are present and accounted for in today's goats.

Above left: *The Lynton goats are descended from stock that was released to live wild in the Valley of the Rocks in north Devon, England, in the late 1800s.*

Above: *Boer goats were developed in South Africa along with Kalahari Red and Savanna meat goats. Today Boers provide much of the goat meat that is produced in North America.*

Bezoar ibex

HEIGHT
Buck: 30–37 in. (76–94 cm);
doe is slightly smaller

WEIGHT
Buck: 100–200 lb. (45–90 kg)
Doe: 90–125 lb. (41–57 kg)

CLASSIFICATION
Ancestor of today's
goats

RELATED BREEDS
All

COUNTRY OF ORIGIN Middle East

Origins and appearance The Bezoar ibex originated in the Fertile Crescent and it still exists in limited numbers in the mountains of Asia Minor, the Middle East, some Aegean Islands, and also on Crete. Bucks are light reddish brown in the summer and ash gray in the winter, with mahogany markings. They have thick beards and scimitar-shaped horns with sharp front edges (keels) and knobs running their length. They're said to have the longest horns of any goat in the world. Does are smaller, with shorter, slender horns that curve gently to the rear. Does remain tawny brown year round.

Behavior and upkeep Bezoar ibex are surefooted and agile, preferring steep, rocky, and forested habitats. They feed in a relatively small area and may be diurnal or nocturnal. Bezoar ibex sometimes climb trees to feed and have been seen 20 ft. (6 m) above the ground. Most of the year, bucks congregate in bachelor herds, while does form family units of a few individuals. As the fall rut approaches, dominant bucks gather harems of 10 to 15 does. The breeding season lasts from November through January, then harems disband for the year. Five months after being bred, does give birth to one to three kids.

Alpine/French Alpine

HEIGHT
Buck: min. 35 in. (89 cm)
Doe: min. 30 in. (76 cm)

WEIGHT
Buck: min. 170 lb. (77 kg)
Doe: min. 130 lb. (59 kg)

CLASSIFICATION
Dairy

RELATED BREEDS
British Alpine,
Mini-Alpine, Oberhasli

COUNTRY OF ORIGIN France

Origins and appearance Originating in the French Alps, Alpines were selected for uniformity, size, and milk production. They are medium to large goats with short hair, straight or slightly dished faces, and narrow, erect ears. They come in an array of colors with French names: *cou blanc* (white front quarters and black hindquarters with black or gray markings on the head), *cou clair* (tan to off-white forequarters shading to gray, with black hindquarters), *cou noir* (black front quarters and white hind quarters), *sundgau* (black with white underbelly and facial stripes), *pied* (splotchy-spotted), and *chamoisée* (brown with specific black facial and leg markings). Any color can be combined ("broken") with large white spots.

Behavior and upkeep Alpines are graceful and curious, although they can be more independent and strong-willed than other dairy breeds. They're hardy and adaptable, so do well in any climate and in both smallholding and large-scale dairying situations. According to American Dairy Goat Association figures, Alpine does produce an average of 2,620 lb. (1,188 kg) of 3.8 percent butterfat milk in a 275- to 305-day period. Only Saanens (see page 189) are more productive. They're persistent milkers, making it possible to milk many goats through for several years before rebreeding.

Valais Blackneck

HEIGHT
Buck: 30–34 in. (76–86 cm)
Doe: 27–31 in. (69–79 cm)

WEIGHT
Buck: 145–200 lb. (66–90 kg)
Doe: 100–35 lb. (45–61 kg)

CLASSIFICATIONS
Dairy, meat, vegetation
management

RELATED BREEDS
None

COUNTRY OF ORIGIN Southern Switzerland, northern Italy

Origins and appearance The Valais Blackneck originated in the Swiss canton of Valais and in the provinces of Verbania and Vercelli in northern Italy. As with many other breeds, populations plummeted in the 1950s when high-production breeds replaced most lower-producing heritage breeds of poultry and livestock, but in the late 1980s, the Valais Cantonal Parliament set aside 1.5 million francs to save it from extinction. This sturdy, bicolored goat has black forequarters and white hindquarters, with a clear demarcation between the halves. It's a medium to large-size goat with coarse, wavy guard hair up to 18 in. (46 cm) long. Animals have short, broad heads, usually with tufts of longer hair on their foreheads. Bucks' massive horns rise up close together then curve up and out near the tips; does' thinner horns curve back and out.

Behavior and upkeep The Valais Blackneck is an agile, sure-footed, hardy mountain breed, able to tolerate wet and cold and to forage its own feed in the summertime. Does produce about 1,100 lb. (500 kg) of milk during a 210-day lactation. Their meat is lean and tasty and usually marketed as capretto (see page 128). Their easygoing dispositions and unusual color also make them popular as estate goats, similar to the Bagot goat (see page 186), and as pets.

West African Dwarf

HEIGHT
Both sexes: 12–20 in.
(30–50 cm)

WEIGHT
Buck: 44–55 lb. (20–25 kg)
Doe: 40–49 lb. (18–22 kg)

CLASSIFICATIONS
Meat, dairy

RELATED BREEDS
American Pygmy,
British Pygmy

COUNTRY OF ORIGIN Africa

Origins and appearance West African Dwarfs, or Djallonké goats, play an important part in the rural economy of West and Central Africa's humid and subhumid zones, where they are kept for milk and meat. The largest population occurs in eastern Nigeria. They play a vital role in food security for rural households, and in times of need they can be sold to provide income. There are two types: humid zone and savanna goats, the latter being somewhat heavier. Both types are achondroplasic dwarves, having wide foreheads and thick, short necks; they are heavy-boned, wide-bodied, and square in appearance. Their color varies, though black is most common, and coat hair is generally short and smooth. Some subtypes have long, wavy hair.

Behavior and upkeep The small size and docile disposition of these goats make them easy to house and handle. They are hardy and fast-maturing, they tolerate tropical heat, and they're trypanotolerant, meaning they resist disease spread by tsetse flies. Because goats of other breeds quickly succumb to trypanosomosis, WADs are the only goats commonly encountered in West and Central Africa. They are also resistant to potentially fatal infestations of barber pole worm (see page 107). Does breed year round and they are prolific, producing two to four kids per kidding.

Nubian/Anglo-Nubian

HEIGHT
Buck: min. 35 in. (89 cm)
Doe: min. 30 in. (76 cm)

WEIGHT
Buck: min. 175 lb. (79 kg)
Doe: min. 135 lb. (61 kg)

CLASSIFICATIONS
Dairy, meat

RELATED BREEDS
Mini-Nubian

COUNTRY OF ORIGIN UK

Origins and appearance British breeders developed today's Nubian during the nineteenth and early twentieth centuries when they crossed native does with long-eared bucks from Asia and North Africa to produce a large dairy goat with pendulous ears, a Roman-nosed profile, and the ability to breed out of season. Nubians carry more flesh than other breeds, so some breeders also market excess kids as meat. A Nubian can be any color, solid or patterned, and its coat is short and silky. The ears lie close to the head and flare out at their rounded tips in a distinctive bell shape; they should extend at least 1 in. (2.5 cm) beyond the muzzle when held flat against the face.

Behavior and upkeep Nubians are docile, affectionate, and become strongly bonded to their owners. Since the typical Nubian's call is loud and strident, with no vibrato like most other breeds, and it often calls to attract its owner's attention, it isn't the best breed to keep in urban situations. Some Nubians are aseasonal breeders and can be bred year round. At about a gallon (3.8 liters) a day, Nubians produce less milk than the other dairy breeds, but what they produce is rich and sweet. According to American Dairy Goat Association figures, Nubian does produce an average of 1,963 lb. (890 kg) of 4.7 percent butterfat milk in a 275- to 305-day period.

Oberhasli

HEIGHT
Buck: min. 30 in. (76 cm)
Doe: min. 28 in. (71 cm)

WEIGHT
Buck: min. 150 lb. (68 kg)
Doe: min. 130 lb. (59 kg)

CLASSIFICATION
Dairy

RELATED BREEDS
Alpine, Mini-Oberhasli

COUNTRY OF ORIGIN Switzerland

Origins and appearance The Oberhasli originated in the mountainous cantons of northern and central Switzerland. They arrived in the US during the early 1900s (initially called Swiss Alpines), and were registered as a subgroup of the French Alpine, but in 1979 the American Dairy Goat Association opened a herdbook and it became a separate breed. The American Livestock Conservancy considers them a recovering breed that still requires monitoring, although they are relatively common in Switzerland, where a herdbook has existed since 1930. Oberhaslis are medium-size, sturdy goats that come in a single color known as *chamoisée*. They are light tan to dark reddish brown, set off by a black belly, boots, forehead, dorsal and facial stripes, and a black/gray udder, although solid black does are also acceptable. Oberhaslis have straight or slightly dished profiles and low-set, upright ears.

Behavior and upkeep Oberhaslis are hardy, friendly, and alert but docile. Because they're unusually strong for their size, wethers make fine harness and packgoats, said to be steadier and less fearful of water than most other breeds. According to American Dairy Goat Association figures, Oberhasli does produce an average of 2,101 lb. (953 kg) of 3.8 percent butterfat milk in a 275- to 305-day period.

Golden Guernsey

HEIGHT
Buck: max. 28 in. (71 cm)
Doe: max. 26 in. (66 cm)

WEIGHT
Buck: min. 150 lb. (58 kg)
Doe: max. 120 lb. (54 kg)

CLASSIFICATION
Dairy

RELATED BREEDS
British Guernsey

COUNTRY OF ORIGIN UK

Origins and appearance Skeletons of goats like Golden Guernseys were unearthed at sites on Guernsey dating to 2000 BCE, so it's thought that goats on Guernsey descend from Syrian and Maltese goats brought there in antiquity— today's examples resemble the Syrian goats with "golden coats" described by Herodotus. Maltese goats still have the same uncommon skin coloring as the Golden Guernsey. Golden Guernseys' coats range in hue from pale blond to deep bronze, and they have orange-red skin. Coat length varies and it's often set off with a fringe of long hair along the spine and down the thighs. About three-quarters of the goats are horned and the rest are naturally polled. Their distinctive erect ears turn back at the tips.

Behavior and upkeep Golden Guernseys produce less milk than other breeds but they do so on forage alone—grass or hay—with little need to supplement this diet with grain. A doe typically produces a quart or two (0.9–1.9 liters) of 3.72 percent butterfat milk per day. Many are precocious milkers that produce milk without first kidding (see page 57) and they often milk through without rebreeding. Golden Guernseys are placid and friendly, and their disposition makes them well suited to homesteading situations where not a great deal of milk is needed.

Icelandic

HEIGHT
Buck: 30–32 in. (76–81 cm)
Doe: 25–28 in. (63–71 cm)

WEIGHT
Buck: 130–65 lb. (59–75 kg)
Doe: 80–110 lb. (36–50 kg)

CLASSIFICATIONS
Dairy, fiber, meat,
vegetation management

RELATED BREEDS
None

Origins and appearance Icelandics, or Settlement Goats, came to Iceland with Norwegian settlers somewhere between 874 and 930 CE. There are no records of further imports since then, so the current population is quite inbred. They are subsidized by the Farmers Association of Iceland to ensure their survival. As a landrace breed, heights and weights vary widely. About 20 percent are white and 80 percent colored—mainly brown, gray, or pied. Horns are of moderate size and curve backward, although a few animals are polled. Icelandics are long-coated with an undercoat of high-quality cashmere. In 1986, six goats were exported to Scotland where they contributed to the development of the Scottish Cashmere goat.

Behavior and upkeep Icelandics are alert, agile, and hardy. They're excellent browsers, ideal for vegetation control. During the coldest months they're stabled and fed hay and supplements, including salted fish—a practice dating back to Iceland's settlement—and the meat from an Icelandic goat is lean and flavorful. Kids mature slowly and dress out (the carcass weight after the head, legs, skin, and internal organs are removed) at about 40 percent live weight. Does produce about 350 to 450 lb. (160–200 kg) of milk during their short lactation, averaging a pint (0.5 liters) per day.

COUNTRY OF ORIGIN Iceland

Dutch Landrace

HEIGHT
Both sexes: max.
25½ in. (65 cm)

WEIGHT
Buck: max. 80 lb. (36 kg)
Doe: max. 60 lb. (27 kg)

CLASSIFICATIONS
Dairy, meat, vegetation
management

RELATED BREEDS
Saanen, Toggenburg

COUNTRY OF ORIGIN The Netherlands

Origins and appearance The Dutch Landrace is one of several long-haired, multicolored, Nordic primitive types kept as landraces in the Netherlands, Iceland, Finland, Sweden, Norway, Denmark, and Britain. It was a primitive landrace type until the early twentieth century, when it was "improved" using Saanen and Toggenburg genetics. The breed was all but extinct by 1958 when the Blijdorp Zoo in Rotterdam initiated a recovery program based on two of the few remaining Dutch Landrace goats; restoration continued elsewhere with an additional four bucks and four does, and then six more animals of unknown ancestry, resulting in today's approximately 2,000 Dutch Landrace. It's a medium-size goat with legs that are short in proportion to its body. It comes in many colors, the most common being brown.

Behavior and upkeep This is a hardy goat that is kept on homesteads as a household dairy and meat goat, but its prime use is as a brush goat on Dutch national reserves, where nearly 1,000 goats in herds of 60 to 120 goats are used to keep areas free of brush and trees. These are active but friendly goats, biddable and easily managed, with tough hooves that need trimming less often than some breeds. They are not agile jumpers and so are fairly easy to contain.

German Improved White

HEIGHT
Buck: 31–35 in. (79–89 cm)
Doe: 27–35 in. (69–89 cm)

WEIGHT
Buck: 165–210 lb. (75–95 kg)
Doe: 120–65 lb. (54–75 kg)

CLASSIFICATION
Dairy

RELATED BREEDS
Saanen

Origins and appearance This North German breed was created in the late 1800s by crossing imported Saanens with native breeds like the local Hessian goat, the now extinct Starkenburg, and white Langensalzas. A herdbook was founded in 1928 when all of Germany's white goat breeds were brought together as German Improved Whites, while all colored goats became German Improved Fawns. German Improved Whites are medium-sized with sturdy, large-framed builds. They have short, smooth, white coats and pale skin, and there can be black spots on the skin but not the hair. Many are polled. They have well-attached, globular udders with medium-long teats suited to both machine and hand milking. They have been shipped around the world where they've been instrumental in the development of several other breeds including, when crossed back to Saanens, the productive Banat White of Romania.

Behavior and upkeep Like their Saanen ancestors, German Improved Whites are calm, friendly, and easily handled; they do well in both homesteading and larger dairy situations. They're also a productive dairy breed, producing 1,550 to 2,200 lb. (700–900 kg) of 3.2 to 3.5 percent butterfat milk in a 300-day lactation. Like other goats of Saanen heritage, they prefer cooler climates and shade in the heat of the day.

COUNTRY OF ORIGIN Germany

Corsican

HEIGHT
Buck: 30–32 in. (76–81 cm)
Doe: 25–28 in. (63–71 cm)

WEIGHT
Buck: 110–35 lb. (50–61 kg)
Doe: 75–90 lb. (34–41 kg)

CLASSIFICATIONS
Dairy, meat

RELATED BREEDS
Alpine

Origins and appearance The origin of indigenous goats on Corsica is uncertain but they've certainly existed there for thousands of years, and French Alpines were used for upgrading in the 1970s. The result is a small to medium-size goat perfectly adapted to life on the island, where they're traditionally herded from mountain pasture to pasture during summer and back to warmer lowlands in fall. Almost 98 percent of the region's goat population is Corsican. Most are long-haired and bearded, with narrow heads and erect, medium-size ears. They have scimitar horns and roughly half have wattles. Corsicans come in an array of colors and patterns. Udder shape varies considerably, though high, globular udders are preferred over long, dangling ones due to potential injury by thorns and rocks.

Behavior and upkeep Corsican does don't give a lot of milk and their lactation periods are short—about 50 gallons (190 liters) per 150-day lactation—but the milk is high in butterfat and milk solids, making it a perfect medium for making farmhouse cheeses. Corsicans are on the move, often with a goatherd, during the summer months, so they're biddable goats that readily accept direction. They do well kept outdoors during the warmer seasons but need shelter from rain and winter storms.

COUNTRY OF ORIGIN Corsica

Damascus

HEIGHT
Buck: 32–34 in. (81–86 cm)
Doe: 27–30 in. (69–76 cm)

WEIGHT
Buck: 170–200 lb. (77–90 kg)
Doe: 140–50 lb. (64–68 kg)

CLASSIFICATIONS
Dairy, show

RELATED BREEDS
None

COUNTRY OF ORIGIN Syria

Origins and appearance Damascus goats, also sometimes called Aleppo, Baladi, Damascene, Halep, or Shami goats, originated in Syria thousands of years ago and are said to be descended from Nubian-type goats. They were exported to Cyprus by the British in the nineteenth century and that's where they're best known today. They're also raised in Lebanon and Syria, where several different subtypes exist. Cyprus-type Damascus goats are elegant creatures with pendulous ears up to 18 in. (46 cm) long, Roman-nosed profiles, small heads, long necks, and long, flowing coats. Shades of brown are the most common color, although black, gray, and goats with red, beige, or white accents occur. Most have light brown eyes, though blue or grey eyes are fairly common. About 60 percent of Damascus goats are naturally polled. Horned does have slender, sickle-shaped horns, while bucks' horns are twisted and spreading.

Behavior and upkeep Damascus goats are intelligent, playful, and easy to handle. Does are good milkers, averaging a little over 3 quarts (around 3 liters) of milk per day. Twins are common. The breed is also greatly appreciated for its appearance, and it is shown at events like the Mazayen al-Maaz competition in Riyadh, Saudi Arabia, where goats compete for the title of "Most Beautiful Goat."

British Alpine

HEIGHT
Buck: 36–38 in. (91–97 cm)
Doe: 30–32 in. (76–81 cm)

WEIGHT
Buck: approx. 170 lb. (77 kg)
Doe: approx. 130 lb. (59 kg)

CLASSIFICATION
Dairy

RELATED BREEDS
Alpine, British Saanen, Mini-Alpine, Nubian, Old English Goat, Saanen, Toggenburg

COUNTRY OF ORIGIN UK

Origins and appearance British Alpines are medium to large goats. They were developed in the UK in the early twentieth century by breeding Alpine bucks from France and Switzerland to native does and to mixed-breed does of Alpine, Toggenburg, Saanen, and Nubian background. A herdbook was established in 1926. They are black with white facial stripes and a white muzzle, white on the ears and rump, and white from the knees and hocks down. Does' coats are short, fine, and glossy, while bucks have somewhat longer hair. British Alpines are tall and somewhat lanky, closely resembling British Saanens in structure. They're noted for their high, well-attached, globular udders.

Behavior and upkeep British Alpines are good-natured, alert, and somewhat independent but friendly and easily handled goats. They're wonderful foragers that don't do well in confinement but thrive on pasture and browse. British Alpines are active goats with finely honed jumping ability so good fences are needed to contain them. Like other goats with Swiss and Alpine backgrounds, they do best in temperate climates. Does produce 1.5 to 2 gallons (5.7–7.6 liters) of roughly 4 percent butterfat milk per day, and many are capable of milking through for at least two or three years.

Toggenburg

HEIGHT
Buck: 34–38 in. (86–97 cm)
Doe: 30–32 in. (76–81 cm)

WEIGHT
Buck: 150–200 lb. (68–90 kg)
Doe: min. 125 lb. (57 kg)

CLASSIFICATION
Dairy

RELATED BREEDS
British Toggenburg

COUNTRY OF ORIGIN Switzerland

Origins and appearance Swiss farmers developed Toggenburgs about 300 years ago in the Toggenburg Valley, and they are said to be the oldest and purest of the Swiss dairy breeds. Toggenburgs, often called Toggs, reached England by 1882, where they became the first recognized dairy breed there. They reached America in 1883 and were subsequently imported in greater numbers than any other breed. These are medium to large goats that come in light fawn to deepest brown but always have the same markings: white ears with a dark spot in the middle of each; white stripes down the face from above each eye to muzzle; hind legs white from hocks to hooves; forelegs white from knees down; white triangles on both sides of the tail; and a white spot at the root of their wattles. Toggs have medium-length coats, straight or slightly dished faces, upright ears, and high, globular, well-attached udders.

Behavior and upkeep Toggs are good-natured, hardy, and alert. Like other Swiss dairy breeds, they do best in cooler climates. Does produce an average of 2,232 lb. (1,012 kg) of 3.1 percent butterfat milk in a 275- to 305-day period. And a Toggenburg doe—GCH Western-Acres Zephyr Rosemary—holds a Guinness World Record title for giving 9,110 lb. (4,132 kg) of milk in 365 days.

Barbari

HEIGHT
Buck: 26–30 in. (66–76 cm)
Doe: 20–24 in. (51–61 cm)

WEIGHT
Buck: approx. 80 lb. (36 kg)
Doe: approx. 50 lb. (22.5 kg)

CLASSIFICATIONS
Meat and dairy

RELATED BREEDS
None

Origins and appearance The Barbari, or Bari, is a small, compact goat distributed throughout the states of Haryana, Punjab, and Uttar Pradesh in India, and Punjab and Sindh provinces in Pakistan; it's also less commonly raised in Mauritius, Nepal, and Vietnam. It originated in the Somalian city of Berbera, and is thought to have been carried to Asia by Somali traders in antiquity. This alert goat has perky, erect, tubular ears, prominent orbital bones, and bulging eyes. The coat is short and sleek, although bucks have large, thick beards. Does' udders are pendulous with large, conical teats. Most Barbaris of both sexes have medium-size, twisted horns that grow up and back. The most common coloring is white overlaid with brownish-red spots.

Behavior and upkeep Barbari goats are a dual-purpose breed raised for meat and milk. The breed is noted for its calm demeanor and suitability for rearing under tethered and stall-feeding conditions. Does are good milkers, producing about a quart of high-butterfat milk per day, over a 150-day lactation. Bucklings are generally castrated at 7 to 30 days of age and then fattened on milk, to be sold for slaughter at Eid and other festivals. Breeding stock is fast-maturing and ready for breeding at an early age, averaging a few months for bucklings, and 8 to 10 months or so for doelings.

COUNTRY OF ORIGIN Somalia

British Saanen

HEIGHT
Buck: min. 32 in. (81 cm)
Doe: min. 30 in. (76 cm)

WEIGHT
Buck: min. 170 lb. (77 kg)
Doe: min. 150 lb. (68 kg)

CLASSIFICATION
Dairy

RELATED BREEDS
Old English Goat,
Saanen

COUNTRY OF ORIGIN UK

Origins and appearance The British Saanen was developed in the early twentieth century by breeding imported Swiss and Dutch Saanen bucks to native or part-Saanen does until a fixed type was achieved and does were at least seven-eighths purebred Saanen. British Saanens look and are colored much like regular Saanens but are longer-legged, longer-headed, and heavier. One of the largest domestic goats ever recorded was a British Saanen named Mostyn Moorcock—400 lb. (180 kg), with a shoulder height of 44 in. (112 cm). British Saanens are strong-boned with deep, wedge-shaped bodies, erect ears, straight profiles, and well-shaped udders. Their parents can be Saanens or British Saanens.

Behavior and upkeep British Saanens are calm, friendly, and docile, tending to get along with their herdmates better than most breeds. Because of that, and the fact that they're high producers, British Saanens are the breed of choice for many intensive dairying operations in the UK. The average British Saanen doe produces in the neighborhood of 1.5 to 2 gallons (5.7–7.6 liters) of 3.5 percent butterfat milk per day. The breed's natural lactation is longer than most, making them ideal for milking through without rebreeding for two or more years.

Kinder

HEIGHT
Buck: approx. 28 in. (71 cm)
Doe: approx. 26 in. (66 cm)

WEIGHT
Buck: 135–50 lb. (61–68 kg)
Doe: 115–25 lb. (52–57 kg)

CLASSIFICATIONS
Dairy, meat

RELATED BREEDS
American Pygmy, Nubian

COUNTRY OF ORIGIN USA

Origins and appearance These dual-purpose goats were originally produced by breeding a NPGA (National Pygmy Goat Association) registered American Pygmy buck to ADGA (American Dairy Goat Association) registered Nubian does, then breeding their first-generation offspring to registered Kinders. They are compact, sturdy, and muscular, with a large body capacity and sound legs; a short, fine coat; a straight or slightly dished face; large, alert eyes; and long, wide ears that rest below the horizontal. All colors and patterns are acceptable. Most Kinders are naturally horned; polled goats are ineligible for registration. A breed association and herdbook was founded in 1988. Kinder is a trademarked breed name.

Behavior and upkeep Kinder does are excellent milkers, averaging about 1,500 lb. (680 kg) of milk (5.5–7.5 percent butterfat content) in 305 days or less. They are aseasonal breeders and breed year round, commonly producing twins, triplets, quadruplets, and even quintuplets. Kids generally weigh 4 to 5 lb. (1.8–2.3 kg) at birth and they mature quickly, reaching slaughter weight of about 50 lb. (23 kg) in six to eight months, and 70 percent of adult growth in one year. Kinders are healthy and hardy, alert but docile, and easily managed.

Mini-Alpine

HEIGHT
Buck: Miniature Dairy Goat
Association: max. 31 in. (79 cm)
Miniature Goat Registry
(purebreds): max. 32 in. (81 cm)

Doe: MDGA: max. 29 in. (74 cm)
MGR: max. 30 in. (76 cm)

WEIGHT
Buck: 100–115 lb.
(45–52 kg)
Doe: 80–100 lb.
(36–45 kg)

CLASSIFICATION
Dairy

RELATED BREEDS
Alpine, Nigerian Dwarf,
West African Dwarf

COUNTRY OF ORIGIN USA

Origins and appearance Mini-Alpines are scaled-down versions of full-size Alpines created, in the first (F1) generation, by breeding a registered Nigerian Dwarf buck to a registered full-size French Alpine doe; after that, offspring are generally bred to Mini-Alpines, although full-size genetics can be re-added to the mix at any time. The Mini-Alpine was created to provide a smaller, more easily handled version of the full-size Alpine. They have large, easily milked teats, produce more milk than Nigerian Dwarfs, and the milk has a higher butterfat content than that of full-size Alpines. Mini-Alpines have long necks, deep but fairly lean bodies, straight or slightly dished faces, and narrow, erect ears. They come in the same six French-named color patterns as full-size Alpines (see page 165) and these colors can be combined with large white spots ("broken"). Either brown or blue eyes are acceptable.

Behavior and upkeep Mini-Alpines are graceful, alert, and active. They're excellent milkers, producing from 2 quarts to a gallon (1.9–3.8 liters) or more of roughly 4–6 percent butterfat content milk per day. Mini-Alpines, like their full-size Alpine counterparts, are skilled jumpers and climbers and so good fencing is required to contain them.

British Pygmy

HEIGHT
Both sexes: 15–20 in.
(40–50 cm)

WEIGHT
West African type weight:
44–55 lb. (20–25 kg)
Southern Sudan type weight:
24–55 lb. (11–25 kg)

CLASSIFICATION
Pet

RELATED BREEDS
American Pygmy,
Nigerian Dwarf,
West African Dwarf

COUNTRY OF ORIGIN UK

Origins and appearance British Pygmy goats are descended from small, hardy, equatorial African dwarf breeds. The British Pygmy Goat Society registers Pygmies exhibiting signs of achondroplasia dwarfism (including disportionately short legs; broad, heavy bodies; and short heads) as well as those afflicted with pituitary hydroplasia (small but normally proportioned dwarf goats) as well as intermediates of both types (see pages 68–69). Their ancestors were exported from Africa to Sweden and Germany as zoo animals in the late 1800s and from there to the UK, the US, and Canada. They come in all colors and color combinations, although light-colored "Swiss stripe" facial markings are forbidden. Hair coats vary in density and length, and bucks have full, long beards and a thick mane of hair over their shoulders.

Behavior and upkeep Pygmies are hardy, quiet, and alert, yet docile, making them ideal pets for children and older folk. They require less space and feed than larger goats and can thrive, where permitted, in the suburbs and on homesteads. Does can be milked, and better British Pygmy milkers yield 1 to 2 quarts (0.9–1.9 liters) of 6 to 10 percent butterfat milk per day. Pygmies are aseasonal breeders that come in heat and can kid year round, producing two to five kids per kidding.

Boer

HEIGHT
Buck: 28–32 in. (71–81 cm)
Doe: 26–28 in. (66–71 cm)

WEIGHT
Buck: 200–340 lb. (90–155 kg)
Doe: 190–230 lb. (85–105 kg)

CLASSIFICATION
Meat

RELATED BREEDS
None

COUNTRY OF ORIGIN South Africa

Origins and appearance Dutch farmers in South Africa's Eastern Cape began developing Boer goats in the 1900s, crossing indigenous goats with Indian and European imports, then selecting for meat rather than dairy production. Boers were initially sent to Australia and New Zealand, then introduced to Europe and the UK in the mid-1980s, and North America in 1993, where they quickly became popular, as they still are today. Boers are heavily muscled, long and wide in the loin, and deep through the body. They have Roman-nosed profiles, pendulous ears, and dark, round horns that curve back before turning out. Their skin is loose and supple; bucks often have rolls of loose skin on their necks and shoulders. The traditional color is white with a red head, neck, and shoulder, but black, red, spotted, and wildly dappled colors occur.

Behavior and upkeep Boer goats are friendly, placid, and easily handled. They're aseasonal breeders and can kid year round. Twins and triplets are the norm and Boer does produce a good supply of high butterfat-content milk, so kids mature quickly. At first, American Boers were prone to kidding problems and heavy nematode infestations (see page 107)—the goats were expensive, so every Boer, faults or not, was used for breeding—but culling has signficantly reduced these issues.

Kiko

HEIGHT
Buck: 30–34 in. (76–86 cm)
Doe: 28–30 in. (71–76 cm)

WEIGHT
Buck: 250–300 lb. (115–135 kg)
Doe: 100–150 lb. (45–68 kg)

CLASSIFICATIONS
Meat, vegetation management

RELATED BREEDS
None

Origins and appearance The Kiko was produced by feral-goat farmers in the 1970s who wanted a hardy meat goat that could produce fast-growing kids (see also page 127). These stocky, muscular goats have medium-length ears that can be erect, stick out to the sides, or flop down. They also have straight profiles; enormous, twisting horns; tough hooves; and short to medium-length coats. Any color is acceptable but most are white. The Kiko's main competitor in North America is the Boer (see page 183). In a 2003/4 study, Tennessee State University measured the performance of 81 Boer, 64 Kiko, and 59 Spanish does. During the study, 71 percent of Boers were treated for lameness versus 31 percent of Kikos, while 50 percent of the Boers required deworming but only 17 percent of the Kikos. The Kikos produced more kids than the Boer and the Spanish does, and their kids weighed more.

Behavior and upkeep Kiko goats are extremely hardy. Their tough hooves don't require trimming as often as other breeds, and their greater resistance to parasites makes them ideal for hot, humid climates where internal nematodes are especially problematic (see page 107). They don't require coddling, either; does kid two or three kids without assistance and Kikos fare better kept outdoors than in.

COUNTRY OF ORIGIN New Zealand

Miniature Silky Fainting Goat

HEIGHT
Buck: ideally max. 23½ in.
(60 cm); acceptable up to
25 in. (64 cm)
Doe: ideally 22½ in. (57 cm);
acceptable up to 23½ in.
(60 cm)

WEIGHT
Buck: 40–75 lb. (18–34 kg)
Doe: 40–60 lb. (18–27 kg)

CLASSIFICATIONS
Novelty, pet, show

RELATED BREEDS
Myotonic, Nigerian
Dwarf

COUNTRY OF ORIGIN USA

Origins and appearance Renee Orr of Lignum, Virginia, developed Miniature Silky Fainting Goats (or Mini-Silkies), in the 1990s by breeding two small, hairy myotonic bucks purchased from veteran fainting goat breeder Frank Baylis to her own long-coated Nigerian Dwarf does, which had myotonic goats in their background. Orr then founded the Miniature Silky Fainting Goat Association in 2005. Mini-Silkies are small goats with soft, flowing coats that sweep the ground, and lots of facial hair, with long bangs and cheek muffs. They come in all caprine colors, including spotted.

Behavior and upkeep Many but not all Mini-Silkies stiffen and sometimes fall over when frightened or excited. This trait, caused by muscular contraction associated with a genetic condition called myotonia congenita (see page 66), is not harmful and fainting goats don't truly faint. The goats remain fully conscious and when their muscles relax, they climb to their feet with no ill effects. This trait renders them more susceptible to predation, so maintaining them within strong, tall fences is a must. Like the other myotonic breeds, Mini-Silkies are known for their friendly, outgoing dispositions. Their lush coats require more grooming than most other breeds.

Bagot

HEIGHT
Both sexes: 23–26 in.
(58.5–66 cm)

WEIGHT
Varies greatly, with no strict
breed standard

CLASSIFICATION
Vegetation management

RELATED BREEDS
None

COUNTRY OF ORIGIN England

Origins and appearance The Bagot goat is an ancient parkland breed kept mainly for its ornamental and historical value. Several theories abound about its origin but none have been proven. What's known is that it was present at Bagot Park, 3 miles from Sir John Bagot's Blithfield Hall in Stafforshire, as early as 1389. It's so associated with the Bagot family that it appears on their coat of arms, and goat head crests are carved on their ancient family tombstones. Bagot goats are small to medium in size, long-haired, and distinctively colored with black forequarters to just behind the shoulders and white hindquarters, sometimes with black patches on their hindquarters and white blazes on their faces. Bucks have huge, sharp-keeled horns up to 36 in. (91 cm) long that sweep back with very little lateral twist; does' horns are smaller.

Behavior and upkeep These self-reliant creatures existed as a single isolated herd under semi-feral conditions for hundreds of years. They're hardy and well adapted to the British climate, being more tolerant of rain than other breeds. They're generally kept outdoors except when kidding. Bagot goats are efficient browsers and excel at conservation browsing, keeping pastures and parklands free of brush and saplings.

Kalahari Red

HEIGHT
Buck: 26–30 in. (66–76 cm)
Doe: 22–26 in. (56–66 cm)

WEIGHT
Buck: approx. 225 lb. (115 kg)
Doe: approx. 165 lb. (75 kg)

CLASSIFICATION
Meat

RELATED BREEDS
None

Origins and appearance The exact origin of Kalahari Red goats is uncertain. They evolved via natural selection in the harsh, arid, and semi-arid desert areas of South Africa where their sun-resistance and hardiness are great perks. Although their conformation is very similar, they are not related to South African Boer goats. Kalahari Reds' deep red coloration and fully pigmented dark skin provides camouflage that helps protect free-ranging herds from predators. They are Roman-nosed, medium in height, and very stocky, with loose skin on their necks. They have long, floppy ears and sturdy, medium-size horns that slope back from the head with a slight outward curve at the tips. Their coats are short and glossy.

Behavior and upkeep These are the quintessential meat goat. Their carcass weight rivals that of the Boer and their flesh is lean, tender, and tasty. Does generally produce twins, though triplets are fairly common. The Kalahari Red is said to be more disease- and parasite-resistant than other South African goats, so requires less maintenance. Because dewormers and vaccines can frequently be dispensed with, kids are often raised to produce organic meat. These are mellow, easily handled goats, but previously unhandled bucks may be aggressive.

COUNTRY OF ORIGIN South Africa

Angora

HEIGHT
Buck: 48 in. (122 cm)
Doe: 36 in. (91 cm)

CLIPPED WEIGHT
Buck: 180–225 lb. (82–102 kg)
Doe: 70–110 lb. (32–50 kg)

CLASSIFICATION
Fiber

RELATED BREEDS
Nigora, Pygora

COUNTRY OF ORIGIN Turkey

Origins and appearance This ancient breed possibly dates back 2,000 years. Spain imported Angora goats from Turkey in the 1700s, as did France, but neither group thrived. However, shipments to South Africa and the US in the 1800s did. Angora goats produce a silky fiber known as mohair (Angora fiber comes from the Angora rabbit), and ideally, fiber is of an equal length throughout the body, from the neck to the rump and belly. Angora goats were originally white; kids that weren't pure white were rigorously culled. Now, with the formation of the Colored Angora Goat Breeders Association, colored Angoras have become popular. They produce black (ranging from deep black to gray and silver), red, and brown fiber, in an array of beautiful patterns.

Behavior and upkeep Angora goats are easygoing and slow-growing—they don't reach mature weight until about 5 years of age. They are also rarely disbudded because the breed is traditionally horned and must have horns in order to be shown. Bucks' heavy, spiral horns sometimes reach 2 ft. (60 cm) or more in length. Does generally give birth to a single kid per year. Angoras thrive best in warm, dry climates; they cannot withstand cold wet rains after shearing. In the US, they were traditionally raised on the Edwards Plateau of west central Texas where the climate is ideal.

Saanen/Sable

HEIGHT
Buck: min. 32 in. (81 cm)
Doe: min. 30 in. (76 cm)

WEIGHT
Buck: min. 165 lb. (75 kg)
Doe: min. 135 lb. (61 kg)

CLASSIFICATION
Dairy

RELATED BREEDS
British Saanen, German
Improved White

COUNTRY OF ORIGIN Switzerland

Origins and appearance Saanens originated in the Saanen Valley of western Switzerland. From the late nineteenth century, they were exported to points around the world and used to develop dozens of dairy and dual-purpose breeds, including British, French, Australian, Yugoslav, and Israeli Saanens; the German Improved White; the Dutch White Goat; the Improved North Russian White; the Bulgarian White; and the Banat White of Romania. According to the Food and Agriculture Organization of the United Nations, purebreds are currently found in over 80 countries. Saanens are large, sturdy yet graceful goats with tan skin and short, fine, white, or cream-colored coats. They sometimes have a fringe of longer hair along their spines and hairy "pantaloons" on their thighs. Sables are a color variation of Saanens and can be any color except white or cream, but they are registered in a separate herdbook.

Behavior and upkeep Saanens and Sables are docile, friendly goats. Saanen does produce an average of 2,765 lb. (1,254 kg) of milk in a 275- to 305-day period, while Sables produce about 2,574 lb. (1,168 kg); the average butterfat content is 3.3 percent for both breeds. These goats are exceptionally hardy, though they can overheat and do best in cooler climates. Due to their pale skin, Saanens require deep shade.

Mini LaMancha

HEIGHT
Buck: Miniature Dairy Goat
Association: max.29 in. (74 cm)
Miniature Goat Registry
(purebreds): 23–29 in.
(58–74 cm)

Doe: MDGA: max. 27 in. (69 cm)
MGR: 23–27 in. (58–69 cm)

WEIGHT
Buck: 90–130 lb.
(41–59 kg)
Doe: 75–120 lb.
(34–54 kg)

CLASSIFICATION
Dairy

RELATED BREEDS
LaMancha, Nigerian
Dwarf, West African
Dwarf

COUNTRY OF ORIGIN USA

Origins and appearance Mini LaManchas, or Mini-Manchas, are smaller versions of full-size LaMancha goats, produced in the first generation (F1) by breeding full-size, registered LaMancha does to registered Nigerian Dwarf bucks and then breeding successive generations to Mini LaManchas, although backcrossing to Nigerian Dwarfs or full-size LaManchas is also acceptable. The LaMancha's most distinguishing feature is its short ears, and only bucks with the "gopher" type of ear can be registered (see page 210). Mini LaManchas are fleshier than most of the other miniature dairy breeds and most have short, sleek summer coats. All colors are acceptable, as are brown or blue eyes. Does have well-attached, globular udders and largish, easy-to-milk teats.

Behavior and upkeep Mini LaManchas are friendly and arguably the most laid-back of the miniature dairy breeds. Most does produce 2 quarts to a gallon (1.9–3.8 liters) of roughly 6 to 7 percent butterfat-content milk per day. They are attentive mothers and two to four kids per kidding is the norm. Larger Mini LaMancha wethers are substantial goats, favored by goat packers for their common sense, strength, and willingness to please—traits that make them ideal driving goats, too.

Nigora

HEIGHT
Buck: min. 19 in. (48 cm);
usually 24–25 in. (61–64 cm)
Doe: min. 18 in. (46 cm);
usually 21–22 in. (53–56 cm)

WEIGHT
Buck: approx. 180 lb. (82 kg)
Doe: approx. 90 lb. (41 kg)

CLASSIFICATIONS
Dairy, fiber

RELATED BREEDS
Angora, Nigerian Dwarf

COUNTRY OF ORIGIN USA

Origins and appearance Nigoras have existed since 1994. They are small to medium-size goats with a dairy-type build, and were created by crossing Angora does (standard white Angoras, colored Angoras, or Navajo Angoras) with Nigerian Dwarf bucks. To be eligible for registration, a Nigora may not be more than 75 percent either Angora or Nigerian Dwarf. Nigoras come in all colors and patterns and their fiber may be a different color than their guard hair. They produce one of three types of fiber. Type A fiber, like mohair from Angora goats, falls in lustrous ringlets about 6 in. (15 cm) long, measuring 28 microns or less. It may be an Angora-like single coat but often contains guard hairs. Type B, the most common type, is cashgora—a mix between mohair and cashmere, characterized by fluffy ringlets with distinguishable guard hairs. Fiber measures 3 to 6 in. (7.6–15 cm) long and averages 24 microns. Type C is cashmere with guard hairs, typically 1¼ in. (3 cm) long, measuring 18.5 microns.

Behavior and upkeep Nigoras are amiable and easily handled. Single and twin kids are the norm, though triplets are not unheard of. Nigora goats producing fiber types A and B must be shorn twice a year; type C producers can be combed, plucked, or shorn. They must not be exposed to cold, wet weather right after shearing.

Swedish Landrace

HEIGHT
Buck: approx. 30 in. (76 cm)
Doe: approx. 25 in. (64 cm)

WEIGHT
Buck: approx. 150 lb. (68 kg)
Doe: approx. 90 lb. (41 kg)

CLASSIFICATION
Dairy

RELATED BREEDS
Saanen (very slightly)

COUNTRY OF ORIGIN Sweden

Origins and appearance Domestic goats were present in Sweden by the Stone Age. Direct descendants are the Swedish Landrace (Svensk Lantras), goats of northern Sweden. Little "improvement" has been attempted beyond the introduction of a few Saanens imported after World War I, so this breed is basically an original Northern European goat (closely related to the Norwegian Landrace goat). Even today, most Swedish Landrace are found in northern Sweden. Being a primitive landrace breed, no standardization has been attempted and sizes and weights vary widely. All are long-coated and come in solid black or white and variegated shades of brown, black, and gray. Swedish Landrace can be polled or horned. Does' slender horns sweep straight back while bucks have massive horns that sweep back, out, and up.

Behavior and upkeep Does are good milkers, producing 2 to 3½ quarts (1.9–3.3 liters) of high-butterfat milk per day, over a 200-day lactation. Swedish Landrace are exceptionally cold-hardy, and excellent browsers and grazers. They adapt well to most management systems, though they were traditionally raised in the mountains except for the coldest months of winter, when they were stabled and fed until spring. These are friendly, docile goats that respond favorably to kind handling.

British Guernsey

HEIGHT
Buck: min. 28 in. (71 cm)
Doe: 26 in. (66 cm)

WEIGHT
Buck: min. 150 lb. (68 kg)
Doe: min. 120 lb. (54 kg)

CLASSIFICATION
Dairy

RELATED BREEDS
British Saanen, Golden
Guernsey, Saanen

COUNTRY OF ORIGIN UK

Origins and appearance Golden Guernseys were exported from the Channel Islands to Great Britain in 1965. Because purebreds were in short supply, British goat fanciers began an upgrading program in which Golden Guernsey does were bred to Saanen or British Saanen bucks, then their female offspring bred back to Golden Guernsey bucks for two generations. Later, Golden Guernsey bucks were bred to does, continuing for a few generations until seven-eights Golden Guernseys were created—these were British Guernseys. Nowadays, either Golden Guernsey or British Guernsey bucks can be bred to British Guernsey does to produce British Guernsey kids. British Guernseys are slightly bigger, slightly heavier-boned versions of the Golden Guernsey with the same basic conformation, coat types, gold color, and reddish-orange skin.

Behavior and upkeep British Guernseys have the same looks and mellow dispositions as their parent breed. Some, however, produce more milk than Golden Guernseys: about 2,200 lb. (1,000 kg) of 3.75 percent butterfat milk per lactation. Friendly, placid, easily handled, and hardy, they are ideal family and homesteading goats, although because they milk well and mostly coexist with peers under intensive conditions without battling, some large commercial dairies in Britain also use them.

Nigerian Dwarf

HEIGHT
Buck: American Dairy Goat
Association/American Goat
Society: 23½ in. (60 cm)
Nigerian Dwarf Goat
Association: max. 23 in. (58 cm)

Doe: ADGA/AGS/NDGA:
22½ in. (57 cm)

WEIGHT
Buck: 60–80 lb. (27–36 kg)

Doe: 50–75 lb.
(23–34 kg)

CLASSIFICATIONS
Dairy, pet

RELATED BREEDS
American Pygmy, British
Pygmy, Mini-Alpine,
Mini LaMancha,
Mini-Nubian, Nigora,
West African Dwarf

COUNTRY OF ORIGIN USA

Origins and appearance The ancestors of today's Nigerian Dwarfs were West African Dwarf goats imported in the 1900s (page 68). These imports were lumped together and called Pygmy goats but fanciers soon noticed there were two types: the blocky, short-legged achondroplasic goats that eventually became American Pygmys, and a slimmer, longer-legged, more proportional type that was used to develop today's Nigerian Dwarf. Nigerian Dwarfs look much like scaled-down full-size dairy goats. They have slim, upright ears that stick out a bit to the sides; straight profiles; and short, straight coats. Primary colors are black, chocolate, and gold, with or without shadings and white spots, but all colors are acceptable, as are brown or blue eyes.

Behavior and upkeep Its small size, sunny disposition, and excellent milking capacity makes the Nigerian Dwarf one of the most popular dairy breeds in the US. Does breed year round and produce two to four or even five kids per breeding. Kids mature rapidly, with many bucklings showing buck behavior at 2 to 3 weeks old, and able to impregnate does by 2 to 4 months. Does produce between 1 quart and 1 gallon (0.9–3.8 liters) or more of 5 to 10 percent butterfat milk per day. Good fencing is needed to keep these small but active and agile goats where they belong.

Mini-Nubian

HEIGHT
Buck: Miniature Dairy
Goat Association:
max. 31 in. (79 cm)
Miniature Goat Registry:
23–31 in. (58–79 cm)

Doe: MDGA:
max. 29 in. (74 cm)
MGR: 21–29 in. (53–74 cm)

WEIGHT
Buck: 100–120 lb.
(45–54 kg)
Doe: 80–105 lb.
(36–48 kg)

CLASSIFICATION
Dairy

RELATED BREEDS
Nigerian Dwarf, Nubian,
West African Dwarf

COUNTRY OF ORIGIN USA

Origins and appearance A first generation (F1) Mini-Nubian is created by breeding a registered Nigerian Dwarf buck to a registered full-size Nubian doe, then breeding their offspring and successive generations to Mini-Nubians, Nigerian Dwarfs, or full-size Nubians. The object of the cross is to produce a medium-size, easily handled goat that resembles a Nubian and that has larger teats and produces more milk than a Nigerian Dwarf. Mini-Nubians are elegant goats with long, wide, pendulous ears and straight or Roman-nosed facial profiles. They have large, deep bodies; long, graceful necks; strong legs; well-attached, globular udders; and supple skin. Any color is acceptable, as are brown or blue eyes.

Behavior and upkeep Mini-Nubians are friendly, love attention, and form strong bonds with handlers. They have loud voices and "talk" more than other breeds, so especially vocal Mini-Nubians may not be suitable for urban locations. They are ideal home milkers that produce about two-thirds the amount of milk as a full-size Nubian on half the feed. Milk production averages 2 quarts to a gallon (1.9–3.8 liters) of 3.5 to 6 percent butterfat milk per day. Some but not all Mini-Nubians are aseasonal breeders, and does typically produce two to four kids per kidding.

British Toggenburg

HEIGHT
Buck: 35–36 in. (89–91 cm)
Doe: 30–33 in. (76–84 cm)

WEIGHT
Buck: 160–220 lb. (73–100 kg)
Doe: 135–60 lb. (61–73 kg)

CLASSIFICATION
Dairy

RELATED BREEDS
Mini-Toggenburg,
Toggenburg

Origins and appearance Breeders in the UK developed British Toggenburgs in the early twentieth century by breeding purebred Toggenburg bucks to native does and does of native and Swiss breed mixes. The breed was opened to crosses in 1925 and closed in 1943; since then they've been considered purebreds. British Toggenburgs are brown, chocolate, or fawn with a white line each side of their faces from above their eyes to their muzzles. They have white edges on their ears, white on their rumps and tails, and their legs are white from the knees down. They have short coats, although bucks' coats are longer and coarser than does'. British Toggenburgs are larger and rangier than Swiss purebreds and they generally give more milk.

Behavior and upkeep British Toggenburgs are alert but docile and easily managed, and they are exceptionally long-lived. Because of their size and high milk production—does produce roughly 2,646 lb. (1,200 kg) of 3.68 percent butterfat milk per lactation—they do require good feed and pasture. They've also been exported to numerous countries around the world but don't fare well in hot, humid climates. William de Goat, the famous World War II goat mascot, was a British Toggenburg wether (see page 149).

COUNTRY OF ORIGIN UK

Appenzell

HEIGHT
Buck: approx. 34 in. (86 cm)
Doe: approx. 30 in. (76 cm)

WEIGHT
Buck: approx. 140–45 lb.
(64–66 kg)
Doe: approx. 100 lb. (45 kg)

CLASSIFICATION
Dairy

RELATED BREEDS
None

Origins and appearance Also known as the Appenzellerzeige (Germany) and Chèvre d'Appenzell (France), the Appenzell originated in the Appenzell region of northwest Switzerland more than 100 years ago. The original Appenzells were white, black, brown, or pied, but in the early 1900s, breeders began selecting for long-haired, polled white goats and the breed remains so today. Large herds once summered in the high mountains; now it's a rare heritage breed subsidized by the Swiss goat breeders association, the Schweizerischer Žiegenzuchtverband. Appenzellers have not been widely exported. This is the smallest of the Swiss breeds. Pure white with pink skin, it resembles the Saanen but is hornless, stockier, hairier, more muscular, and less refined. H.S. Holmes Pegler's *Book of the Goat* (1886), claimed the Appenzell was "practically a Toggenburg," being the same type and differing only in color.

Behavior and upkeep Appenzells are docile, playful, and friendly. They are also very hardy and can tolerate most climates: they thrive on rocky mountain pasture, although they require stabling during the coldest months. A typical doe produces a significant quantity of milk—1,650 to 1,765 lb. (750–800 kg) of 2.9–3.0 percent butterfat milk in a 270-day lactation—and much of this is used to craft cheese.

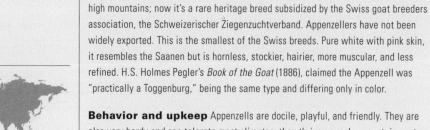

COUNTRY OF ORIGIN Switzerland

Old English Goat

HEIGHT
Buck: ideally 28 in. (71 cm);
acceptable 26–30 in.
(66–76 cm)
Doe: ideally 26 in. (66 cm);
acceptable 24–28 in.
(61–71 cm)

WEIGHT
Both sexes: 100–120 lb.
(45–55 kg)

CLASSIFICATIONS
Rare, heritage

RELATED BREEDS
Arapawa, British Feral
Goat, Old Irish Goat

COUNTRY OF ORIGIN UK

Origins and appearance Old English Goats descend from stock brought to Europe by Neolithic farmers, evolving as hardy goats that thrived in a harsh climate and on marginal land while producing tasty kids and some milk. They are small and stocky with deep, broad bodies and short, sturdy legs. They have moderately fine muzzles, dished faces, and small, pricked ears. Both sexes are bearded and most are horned, though polled individuals are not unheard of. Horns rise straight from the head and then curve back or twist outward. Old English Goats come in most colors including spotted, though gray or brown—with or without black markings—is most commonly seen. Markings indicative of Swiss dairy goat lineage are unacceptable. The outer coat can be short or shaggy but never smooth or sleek, over an undercoat of fine cashmere. Goats are registered with the Old English Goat Society, and redomesticated British feral goats are considered part of the breed.

Behavior and upkeep These goats are alert and active. They thrive on browse, making them ideal candidates for conservation grazing. The homesteader's goat of earlier times, it's still worth milking when high production isn't a prime requisite. It also yields tasty meat and shaggy pelts. Bucks' horned skulls also achieve high prices.

American Pygmy

HEIGHT
Buck: max. 23½ in (60 cm)
Doe: max. 22½ in (57 cm)

WEIGHT
Buck: 60–90 lb. (27–41 kg)
Doe: 50–80 lb. (23–36 kg)

CLASSIFICATIONS
Pet, dairy

RELATED BREEDS
British Pygmy, Nigerian Dwarf, West African Dwarf

COUNTRY OF ORIGIN USA

Origins and appearance American Pygmys are direct descendants of West African Dwarf goats. The first documented importation of West African Dwarf and South Sudanese goats from Africa occurred in 1909 and more arrived between the 1930s and late 1960s. Fanciers then formed the National Pygmy Goat Association in 1975. Pygmy goats have erect ears, dished faces, and full coats of straight, medium-length hair. They're genetically small, cobby, compact, and muscular, with limbs and head that are short in relation to their body width and length. Pygmys come in solid black; black with white accents; black, brown, and gray agouti (grizzled black base coat with black or brown accents); and black or brown caramel (white to dark tan, accented with black or brown).

Behavior and upkeep Friendly, intelligent, and docile, American Pygmys are popular as pets. Does produce from 1 to 2 quarts (950–1,890 ml) of milk with 5 to 10 percent butterfat content that is said to be higher in calcium, phosphorus, potassium, and iron than milk from full-size dairy breeds. Does are attentive mothers that breed year round. They produce two to four—and occasionally five—fast-maturing kids per breeding, but to prevent kidding problems, they must not be allowed to become obese.

Pygora

HEIGHT
Buck: min. 23 in. (58 cm)
Doe: min. 18 in. (46 cm)

WEIGHT
Buck: 75–140 lb. (34–64 kg)
Doe: 80–120 lb. (36–54 kg)

CLASSIFICATION
Fiber

RELATED BREEDS
American Pygmy, Angora

COUNTRY OF ORIGIN USA

Origins and appearance Katharine Jorgensen developed Pygoras (medium-size fiber goats with colored fiber) in the 1980s by breeding Pygmy does to white Angora bucks. First generation (F1) crosses from breeding Pygmys to Angoras are not Pygoras but are eligible for registration in the F1 division of the herdbook. F1s must be bred to Pygoras to produce Pygora offspring but their offspring and successive generations can be bred to white Angoras, Pygmys, or Pygoras as long as offspring aren't more than 75 percent Pygmy or Angora. Pygoras can be any Pygmy color and their dilutions, as well as white. They produce one of three kinds of fiber. Type A is similar to fine kid mohair and reaches about 6 in. (15 cm) long; type B is soft, curly, 3 to 6 in. (7.6–15 cm), with characteristics of both mohair and cashmere (cashgora); type C is ultra-soft, matte-finish, and cashmere-like, 1 to 3 in. (2.5–7.6 cm) long. Type A and B goats are shorn twice a year; type C animals can be brushed, plucked, or shorn.

Behavior and upkeep Pygora goats are friendly, playful, and easily handled. They're hardier than purebred Angoras but require good shelter from rain and storms, especially during the weeks after shearing. Pygoras make affectionate pets. They are also efficient browsers, which makes them popular for vegetation management.

Myotonic

HEIGHT
Buck: min. 28 in. (71 cm)
Doe: min. 26 in. (66 cm)

WEIGHT
Buck: min. 150 lb. (68 kg)
Doe: min.120 lb. (54 kg)

CLASSIFICATIONS
Heritage, meat, pet

RELATED BREEDS
Miniature Silky
Fainting Goat

COUNTRY OF ORIGIN USA

Origins and appearance Myotonics, also known as fainting, nervous, or wooden-leg goats, are a landrace breed that originated in Tenneesee in the late 1800s (see page 66). They are muscular and wide in proportion to their height, and due to frequent muscle tensing caused by a genetic condition called myotonia congenita (which causes them to stiffen when excited), their hindquarters are especially well developed. Most are horned, ranging from large, twisted horns to simple, swept-back versions, but some are polled. The average myotonic goat is short-haired, but some have longer, thicker coats, and they come in all colors, patterns, and markings, although black and white is most common.

Behavior and upkeep Myotonic goats are easy to keep. They are friendly and exceptionally docile. They're also more parasite-resistant than many other breeds, and are also kept inside fences more easily (climbing a fence causes them to stiffen). They generally breed year-round, and twin and triplet births are the norm. The degree of stiffness displayed can range from very occasional stiffness without falling, right up to goats that always move stiffly and readily seize up when startled or facing a low barrier.

San Clemente Island

HEIGHT
Buck: min. 27 in. (69 cm)
Doe: min. 26 in. (66 cm)

WEIGHT
Buck: 125–50 lb. (57–68 kg)
Doe: 50–80 lb. (23–36 kg)

CLASSIFICATION
Vegetation management

RELATED BREEDS
None

Origins and appearance This critically endangered breed is monitored by the Livestock Conservancy. It came to San Clemente, off the coast of California, in 1875, carried there by a shepherd from nearby Santa Catalina Island. A feral population thrived until the mid-1980s, when the US Navy ordered their extermination. The Fund for Animals saved the breed by relocating over 3,000 goats to the mainland—most dispersed, but a few breeders gathered small herds and founded the San Clemente Island Goat Foundation. Once thought to be of Spanish origin, DNA testing in 2007 showed that they're unrelated to Spanish goats. San Clemente Island goats are light brown to dark red with black breed-specific markings. They are fine-boned, deer-like, with dished profiles; long, lean faces; and narrow, horizontally carried ears. Bucks have large, outward-twisting horns; does' horns are smaller, curving straight back.

Behavior and upkeep These goats are hardy, alert, active, and agile. They have no commercial value but are excellent browsers, making them ideal for brush control. They're classified as Critical by the Livestock Conservancy, making them attractive for breeders wanting to raise hardy, beautiful little goats, while helping preserve history. However, due to their rarity, breeding stock can be hard to find.

COUNTRY OF ORIGIN USA

Spanish

HEIGHT
Both sexes: variable depending on bloodline

WEIGHT
Both sexes: 50–200 lb. (23–90 kg)

CLASSIFICATIONS
Meat, vegetation management

RELATED BREEDS
None

COUNTRY OF ORIGIN USA

Origins and appearance When Columbus's fleet sailed from Spain in 1492, it carried everything needed to colonize the New World, including goats. Meanwhile Spanish sailors salted islands along nautical routes with goats, knowing they'd survive, multiply, and provide fresh meat on subsequent trips. These became the ancestors of today's Spanish goats. This landrace breed was shaped by natural selection and geographic isolation, so Spanish goats vary from one region (and even one farm) to another. Heads are usually straight or slightly dished with moderately long ears carried alongside the face instead of out to the sides. They have substantial horns and their bodies are somewhat rangy and large-framed. Most have short coats. All colors are acceptable. They are monitored by the Livestock Conservancy.

Behavior and upkeep Spanish goats are alert, hardy, and more parasite-resistant than many breeds. They're avid browsers and thrive under minimal-input conditions, making them very effective brush goats, but they do require sufficient shelter from rain and winter storms. Does are long-lived and prolific, producing one to three kids per kidding. Wethers are friendly and sturdy, and are often used as packgoats and cart goats.

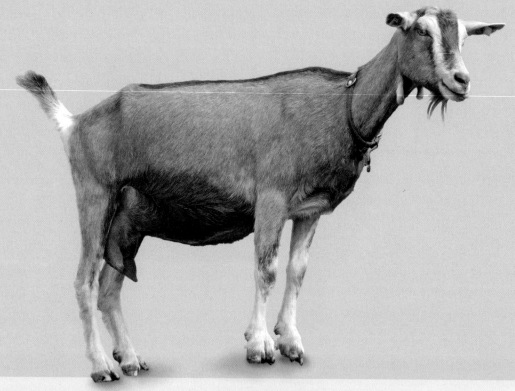

Thuringian

HEIGHT
Buck: 30–32 in. (76–81 cm)
Doe: 29–30 in. (74–76 cm)

WEIGHT
Buck: approx. 125 lb. (57 kg)
Doe: approx. 105 lb. (48 kg)

CLASSIFICATIONS
Dairy, meat, vegetation
management

RELATED BREEDS
Toggenburg

COUNTRY OF ORIGIN Germany

Origins and appearance Also called the Thuringian Forest Goat, Thüringer Waldziege, Thüringer Wald, and the German Toggenburg, the Turingian is a sub-breed of the German Improved Fawn, created in the late 1800s and early 1900s by crossing Swiss Toggenburg, Harzerziege, and Thüringer Landziege goats with native landrace does and selecting for dairy quality, hardiness, and vigor. A herdbook was established in 1935 but the breed is relatively rare today. Thuringians have black, chocolate, or grayish-brown coats with pale or white "Swiss" facial stripes, upright white ears, white lower legs, and a white flash on their rumps. They come in both horned and polled versions. Coats are short to medium in length and close-lying.

Behavior and upkeep Thuringians are agile, sure-footed, hardy, and vigorous, and well suited to mountain terrain. They're relatively parasite-resistant and efficient browsers that, except in late pregnancy or during the deep wintertime, require little in the way of supplementary feeding, so they are often used for brush control and field maintenance. When slaughtered, they produce lean carcasses with a high yield. Does produce about 1,500 to 2,200 lb. (680–1,000 kg) of roughly 3.5 percent butterfat milk per lactation. Twins are the norm, though triplets occur.

Carpathian

HEIGHT
Buck: approx. 27–28 in.
(69–71 cm)
Doe: approx. 24–25 in.
(61–64 cm)

WEIGHT
Buck: approx. 115 lb. (52 kg)
Doe: approx. 80 lb. (36 kg)

CLASSIFICATIONS
Dairy, meat

RELATED BREEDS
None

Origins and appearance The Carpathian goat, also known as the Karpatska or Karpatina, originated during the nineteenth century in the Carpathian Mountains of Poland. Although few remained by the late twentieth century, a small herd was located in 2005 and taken to the National Research Institute of Animal Production where it was brought back from the brink of extinction. It's now found in Poland and Romania. These are long-haired goats with somewhat slim bodies, sloping hindquarters, long necks, straight profiles, and long, narrow ears. Bucks have sturdy, twisted horns; does' horns are thin and sweep straight back. Most Carpathians in Poland have white coats; Romanian goats can be white, red, black, gray, or spotted. During the winter months they grow a fine cashmere undercoat measuring 18 to 24 microns (see page 132).

Behavior and upkeep Carpathian goats are easygoing, exceptionally hardy, and excellent browsers. Does are only modest milkers, producing about a quart or so (a liter) per day over a nine-month lactation, but the milk contains 4.5 to 5 percent butterfat, making it a perfect medium for crafting cheese. Kids are often raised for meat and they are fairly fast maturing, weighing 18 to 30 lb. (8–14 kg) at the slaughtering age of two months. Single kids and twins are the norm.

COUNTRY OF ORIGIN Poland

American Cashmere

HEIGHT
Both sexes: variable

WEIGHT
Both sexes: variable

CLASSIFICATION
Fiber

RELATED BREEDS
None

Origins and appearance All goats, except single-coated breeds like Angoras, grow warm winter underwear called cashmere. Cashmere goats—goats that produce cashmere in commercial quality and quantity—were long considered a type instead of a breed. Breeders, however, are now banding together to create and register American Cashmere goats. According to the Cashmere Goat Registry, these can be any height but must be well balanced, long, deep, and wide, with straight, level backs and sturdy legs. They must be horned; polled animals are ineligible for registration. The most desirable color is white but all other colors are acceptable, although goats should be one solid color on the shearable parts of their bodies (neck, shoulders, sides, hips).

Behavior and upkeep Cashmere goats can be shorn or have their cashmere combed out as it starts to shed. They are alert and active but easily handled. Does are good mothers that rarely require assistance giving birth. Cashmere goats are generally easygoing—a positive trait, especially for Cashmeres whose owners comb them. As with all fiber goats, it's important to keep vegetable matter—manure, sticks, hay, and the like—out of a Cashmere's fleece by maintaining it in clean surroundings and feeding from hay racks and grain pans that don't drop debris on its coat.

COUNTRY OF ORIGIN USA

Old Irish Goat

HEIGHT
Buck: 24–27 in. (61–69 cm)
Doe: 22–24 in. (56–61 cm)

WEIGHT
Both sexes: 100–120 lb.
(45–54 kg)

CLASSIFICATIONS
Heritage, vegetation
management

RELATED BREEDS
Old English Goat

COUNTRY OF ORIGIN Ireland

Origins and appearance The Old Irish Goat is a feral breed that has been shaped by Ireland's climate and landscape since it first arrived with Neolithic settlers 5,000 years ago. Once the primary farmstead goat of Ireland, it's now mainly represented by feral goats at large in mountainous areas across most counties. Old Irish Goats are typical of the Northern Breed Group that includes Old English Goats, Icelandics, and Lappgetters. They are small, stocky goats with short, strong legs and deep bodies, with rounded bellies. They have small ears, dished profiles, and prominent frontal bones. Bucks' horns are immense and strongly keeled; horns sweep back and out in both sexes. Polled individuals are occasionally seen. Coats are long and come in an array of colors.

Behavior and upkeep The Old Irish Goat, when returned to domestication, is a low-maintenance homesteader's goat that produces small quantities of rich milk as well as tasty meat. These goats are hardy and alert but amiable and easily managed. They're also suited for outdoor living except in the coldest winter months, thus perfect for vegetation management. Housing needs are minimal as they prefer being outdoors, but shelter is needed from rain and winter storms.

Mongolian Cashmere

HEIGHT
Both sexes: varies

WEIGHT
Both sexes: varies

CLASSIFICATION
Fiber

RELATED BREEDS
None

COUNTRY OF ORIGIN Mongolia

Origins and appearance Mongolia is one of the world's major producers of cashmere, and this comes from several fiber goat breeds. The breed known simply as the Mongolian goat thrives on dry, cold uplands, facing temperatures from −27° up to 99°F (−33°C to 37°C). It yields white cashmere, though its long, shaggy guard hair can be white, black, blue, brown, or pied. Like the other Mongolian Cashmere breeds, it's lightweight, strongly horned, and small (bucks average 26 in./66 cm in height). A brown Cashmere goat, the Uuliin Bor is raised in western Mongolia. It was developed by crossing Mongolian goats with dark gray, cashmere-producing Altai Mountain goats from Russia. Other important cashmere producers include the Gobi Wool, a dark gray goat developed by breeding Mongolian goats to Don goats from Russia; and the white Atlas Down and Alashan Down goats of Inner Mongolia.

Behavior and upkeep These goats are incredibly hardy. Their hides and guard hair are marketed in addition to their cashmere, and they're also used for meat and household dairy production. They are all cold-weather goats that thrive in the far north, so housing can be minimal; they simply require shelter from wind and driving snow. Temperament varies, but they are generally amiable and easily handled.

Arapawa

HEIGHT
Both sexes: 26–30 in.
(66–76 cm)

WEIGHT
Buck: approx. 125 lb. (57 kg)
Doe: 60–80 lb. (27–36 kg)

CLASSIFICATION
Heritage

RELATED BREEDS
Old English Goat

COUNTRY OF ORIGIN New Zealand

Origins and appearance Ancestors of today's Arapawa goats arrived in New Zealand with explorers, whalers, sealers, and settlers. These were likely Old English Goats, a breed that all but disappeared back in Britain as improved dairy breeds from France and Switzerland were added to the mix. In isolation on Arapawa Island, the old breed bred true and today's goats are said to be a last living remnant of the very original Old English Goat. In the 1980s, the New Zealand Forest Service culled the goats on Arapawa Island, so residents Betty and Walt Rowe set up a sanctuary for 40 goats on their farm. All of today's goats descend from that group and from nine Arapawas recovered from feral herds. Arapawa goats were exported to the US in 1993 and the UK in 2004, but they are endangered worldwide. Goats are tan, brown, white, and black in varying combinations, with black stripes on their faces. Tiny ears, flowing coats, and enormous horns on bucks are hallmarks of the breed.

Behavior and upkeep Arapawa goats are alert and active but easily handled. As with many formerly feral populations, they're exceptionally hardy. They're friendly goats and tend to form close bonds with family groups and humans. Arapawa goats are excellent jumpers so they require good fences to contain them.

LaMancha

HEIGHT
Buck: min. 30 in. (76 cm)
Doe: min. 28 in. (71 cm)

WEIGHT
Buck: min. 155 lb. (70 kg)
Doe: min. 125 lb. (57 kg)

CLASSIFICATION
Dairy

RELATED BREEDS
Mini LaMancha

COUNTRY OF ORIGIN USA

Origins and appearance Developed in the USA in the 1930s, LaManchas are said to descend in part from short-eared goats imported to California's western missions and ranches during the Spanish colonization era. The most distinctive feature of this medium-size goat is its short external ears. "Gopher" ears can be no more than an inch (2.5 cm) long, with little or no cartilage; they often appear shriveled. "Elf" ears are upright and perky. They can be a maximum of 2 in. (5 cm) long and cartilage shapes the ear. With both types, the end of the ear must turn up or down. Only gopher-eared bucks are eligible for registration with the American Dairy Goat Association because two elf-eared individuals bred together can produce offspring with normal erect ears. LaManchas' coats are short, fine, and glossy, and any pattern, color, or color combination is acceptable.

Behavior and upkeep LaManchas tend to be friendly, docile, and even-tempered. They are also hardy and adapt well to any climate. They're easily handled and managed, too, making them well suited to both large-scale and homesteading situations. LaMancha does produce an average of 2,349 lb. (1,065 kg) of 3.7 percent butterfat milk in a 275- to 305-day period.

Lappgetter

HEIGHT
Both sexes: approx.
28 in. (71 cm)

WEIGHT
Both sexes: 50–155 lb.
(23–70 kg)

CLASSIFICATION
Dairy

RELATED BREEDS
None

COUNTRY OF ORIGIN Sweden

Origins and appearance Also known as the Sampi, Lappget, or simply Lapp goat, this breed is native to northern Sweden, where it was raised by nomadic Sami reindeer herders for milk and meat. The herders also used its skin to sew bags, hats, and other garments, and its horns for crafting buttons, eating utensils, and gunpowder horns. Thought to be extinct, in 2001 seven goats were found near Fatomakke in South Lapland. The Swedish government has taken steps to preserve this heritage breed and though still critically endangered, it's staging a comeback. Being a landrace breed, heights and weights vary widely. Lappgetters are somewhat short-legged in proportion to body size, with a broader forehead than the norm. Most are white, but black, black with gray, and spotted individuals occur. Bucks' sturdy horns sweep back and sometimes turn slightly out at the tips; does have slender, swept-back horns.

Behavior and upkeep Lappgetter goats are alert but amiable and fully adapted to life in the mountainous far north, so they're very winter-hardy. They do, however, need shelter from storms and extreme cold. They're efficient foragers and happily go out in cold weather and deep snow to browse. Lappgetters can be milked, and though production is small, their milk is high in milk solids, making it ideal for crafting cheese.

Appendices

Glossary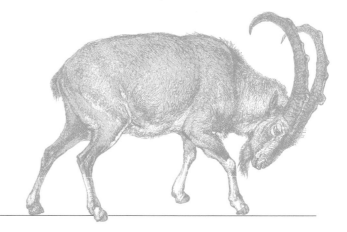

ABOMASUM—The third compartment of the ruminant stomach; the compartment where digestion takes place

AMMONIUM CHLORIDE—A mineral salt fed to male goats to inhibit the formation of bladder and kidney stones

ANESTRUS—The period of time when a doe is not having estrous (heat) cycles

ASEASONAL BREEDER—A doe that comes in heat year round; goat breeds that originated fairly close to the Equator, including Boers (South Africa), Kikos (New Zealand), and West African Dwarfs, are aseasonal breeders

BILLY—(slang) An uncastrated male goat; the preferred term is "buck"

BLEATING—Goat vocalization; also referred to as calling

BOLUS—A large, oval pill; also used to describe a chunk of cud

BREECH BIRTH—A birth in which the rump of the kid is presented first

BREED—Goats of a color, body shape, and other characteristics similar to those of their ancestors, capable of transmitting these characteristics to their own offspring

BROKEN-MOUTH—A goat that has lost some of its permanent incisors, usually at five or more years of age

BROWSE—Morsels of woody plants including twigs, shoots, and leaves

BUCK—An uncastrated male goat

BUCKLING—An immature, uncastrated male goat; an uncastrated male kid

BUNTING—The act of a kid poking its dam's udder to stimulate milk let-down

BUTTING—The act of a goat bashing another goat (or a human) with its horns or forehead

CABRITO/CAPRETTO—Spanish: "baby goat"; meat from a milk-fed kid

CALLING—Goat vocalization; also referred to as bleating

CAPRINE—Having to do with goats

CHEVON—Goat meat

COBBY—Chunky, heavily built, but with short legs

COLOSTRUM—First milk a doe gives after birth; high in antibodies, this milk protects newborn kids against disease; sometimes incorrectly called colostrums

CONCENTRATE—High-energy, low-fiber, highly digestible feed such as grain

CROSSBREED—An animal resulting from the mating of two entirely different breeds

CUD—Undigested food regurgitated by a ruminant to be chewed and swallowed again

DAM—The female parent

DENTAL PAD—An extension of the gums on the front part of the upper jaw; it is a substitute for top front teeth

DEHORNING—The removal of existing horns

DISBUD—To destroy the emerging horn buds of a kid via the application of a red-hot disbudding iron

ESTRUS—The period when a doe is receptive (will mate with a buck; e.g. she is "in heat") and can become pregnant

ESTRUS CYCLE—The doe's reproductive cycle

FAINTING GOAT—A common name for myotonic goats

FLEHMEN—Curling of the upper lip in order to increase the ability to discern scent

FORAGE—Grass and the edible parts of browse plants that can be used to feed livestock

FOUNDATION ANIMALS—The first animals used in creating a new breed; the first animals used in a specific breeding program

GRAIN—Seeds of cereal crops such as oats, corn, barley, milo, and wheat

HERMAPHRODITE—An animal with both male and female sex organs

LACTATION—The period when a doe is giving milk

LANDRACE—A locally adapted, traditional breed or variety of a species developed in isolation over time

MASTITIS—Inflammation of the udder

MICRON—A unit of length equal to one millionth of a meter and used to indicate the diameter of a hair shaft

MILK LET-DOWN—The release of milk by the mammary glands

MONKEY MOUTH—See underbite

MYOTONIA CONGENITA—The inherited neuro-muscular condition that causes myotonic goats' major muscles to temporarily seize up

MYOTONIC—The preferred name for "fainting goat"; a goat carrying the gene for myotonia congenita

NANNY—(slang) A female goat; the preferred term is "doe"

OMASUM—The third part of the ruminant stomach; the omasum is sandwiched between the reticulum and the abomasum

OVERSHOT/PARROT MOUTH—When the lower jaw is shorter than the upper jaw and the teeth hit in back of the dental pad

PHENOTYPE—An individual's observable physical characteristics

PIZZLE—The urethral process, a stringy-looking structure at the end of a male goat's penis

POLLED—A natural absence of horns

POSTPARTUM—After giving birth

PREPARTUM—Before giving birth

PROLIFIC—Producing more than the usual number of offspring

PUREBRED—An animal of a recognized breed that is kept pure for many generations

RATION—Total feed given to an animal during a 24-hour period

REGISTERED ANIMAL—An animal that has a registration certificate and number issued by a breed association

RETICULUM—The second chamber of a ruminant's stomach

ROMAN-NOSED—The convex profile of breeds like the Boer meat goat and Nubian dairy goat

RUMEN—The first compartment of the stomach of a ruminant, in which microbes break down cellulose

RUMINANT—An animal with a multi-compartment stomach

RUMINATION—The process whereby a cud or bolus of rumen contents is regurgitated, re-chewed, and re-swallowed; "chewing the cud"

RUT—The period during which a buck is interested in breeding females

SEASONAL BREEDERS—Does that only come in heat during part of the year; most dairy goats are seasonal breeders

SELECTION—Choosing superior animals as the parents of future generations

SLAUGHTER KID—A kid produced specifically for the meat market

SMOOTH MOUTH—A goat that has lost all of its permanent incisors, usually seven or more years of age

SOW MOUTH—See **UNDERBITE**

STANDING HEAT—The period during estrus (heat) when a doe allows a buck to breed her

URINARY CALCULI—Mineral salt crystals ("stones") that form in the urinary tract and sometimes block the urethras of male goats

UDDER—The female mammary system

UNDERBITE—The lower jaw is longer than the upper, and the teeth extend forward past the dental pad on the upper jaw; also known as monkey mouth or sow mouth

URETHRAL PROCESS—The pizzle; a stringy-looking structure at the end of a male goat's penis

WETHER—A castrated male goat

YEARLING—A goat of either sex that is one to two years of age, or a goat that has cut its first set of incisors

Bibliography

BOOKS

Baldwin, H. (1918) *Holding the Line.* A.C. McClurg, Chicago.

Bewick, T. (1800) *A General History of Quadrupeds.* Hodgson, Belby & Bewick, Newcastle, UK.

Boldrick, L. (1996) *Pygmy Goats: Management and Veterinary Care.* All Publishing Company, Orange, California.

Caldwell, G. (2017) *Holistic Goat Care.* Chelsea Green Publishing, White River Junction, Vermont.

Clutton-Brock, J. (1999) *A Natural History of Domesticated Mammals.* Cambridge University Press, Cambridge, UK.

Cooper, T. (2018) *Goat Behavior: A Collection of Articles.* Karmadillo Press, Cheshire, Oregon.

Epstein, H., and Mason, I. (1971) *The Origin of the Domestic Animals of Africa.* Africana Publishing Corporation, New York.

Hall, A. (1982) *The Pygmy Goat in America.* Hall Press, San Bernardino, California.

Hinson, J. (2015) *Goat.* Reaktion Books, London.

Holmes-Pegler, H. S. (1909) *The Book of the Goat* (4th ed.). L. Upcott Gill, London.

Houpt, K. (2011) *Domestic Animal Behavior for Veterinarians and Animal Scientists* (5th ed.). Wiley-Blackwell, Hoboken, New Jersey.

Hurst, J. (2014) *Extraordinary Goats: Meetings with Remarkable Goats, Caprine Wonders & Horned Troublemakers.* Voyageur Press, Minneapolis.

Hussain, M. S. (2010) *Essentials of Caprine Anatomy: A Complete Text on General and Systemic Anatomy.* University of Agriculture, Faisalabad, Pakistan.

Jensen, P. (2009) *The Ethology of Domestic Animals: An Introductory Text.* Association of International Research and Development Centers for Agriculture, Wallingford, UK.

Mackenzie, D. (1993) *Goat Husbandry.* Faber and Faber, London/New York.

Naaktgeboren, C. (2006) *The Mysterious Goat: Images and Impressions.* BB Press, Eindhoven, Netherlands.

Porter, V. (1996) *Goats of the World.* Farming Press, Ipswich, UK.

Shields, J. (1982) *Exhibition and Practical Goatkeeping.* Triplegate Ltd., Hindhead, Surrey, UK.

Smith, C. (2019) *Goat Health Care* (5th ed.). Karmadillo Press, Cheshire, Oregon.

Stockwell, F. (2014) *Beautiful Goats: Portraits of Classic Breeds.* Ivy Press, Lewes, UK.

Thear, K. (1988) *Goats and Goatkeeping.* Merehurst Press, London.

Waite, B. (2008) *William de Goat: The Story of Air Commodore William de Goat DSO DFC, the Extraordinary Mascot of 609 (West Riding) Squadron during the Second World War.* Athena Press, Twickenham, UK.

BULLETINS

European Regional Conference on Goats (2014) "Sustainable Goat Breeding and Goat Farming in Central and Eastern European Countries." Food and Agricultural Organization of the United Nations. (fao.org/3/a-i5437e.pdf)

Fernandez, D. "Introduction to Goat Reproduction." Cooperative Extension Program, University of Arkansas at Pine Bluff. (uapb.edu/sites/www/Uploads/SAFHS/FSA-9607.pdf)

New South Wales Department of Primary Industries (2017) "Anatomy and Physiology of the Goat." (dpi.nsw.gov.au/__data/assets/pdf_file/0010/178336/Anatomy-and-physiology-of-the-goat.pdf)

Onderstepoort Veterinary Institute with KwaZulu-Natal Department of Agriculture and Environmental Affairs (2007) "Goatkeepers' Animal Health Care Manual," 2nd edition. (lib.icimod.org/record/13031/files/1509.pdf)

Savage, Sara (2013) "Goat Enrichment." Winter 2013 issue of the Enrichment Record. (enrichmentrecord.com/wp-content/uploads/2011/01/goat-enrichment.pdf)

Sponenberg, P, and Roberts, B. (2005) "Myotonic Goat Description." (myotonicgoatregistry.net/MGRbreeddescription/MGRBreeddescription.html)

Stanton, T. (2012) "Kidding with Confidence." Cornell Sheep & Goat Extension Program, Cornell University. (putknowledgetowork.org/resources/kidding-with-confidence)

Trinidad and Tobago Goat and Sheep Society, St. Augustine, Trinidad. "The Trinidad and Tobago Dairy Goat Manual: Breeds, Milking, Herd Health, Records." (infoagro.net/sites/default/files/migrated_documents/attachment/EN_TTGSS_Goat_Manual.co.pdf)

USDA Sustainable Agriculture Research and Education Program, University of Rhode Island (2016) "How and Why to Do FAMACHA Testing." (web.uri.edu/sheepngoat/files/FAMACHA-Scoring_Final2.pdf)

ARTICLES

Adedeji, T. A. et al. (2012) "Effect of Some Qualitative Traits and Non-Genetic Factors on Heat Tolerance Attributes of Extensively Reared West African Dwarf (WAD) Goats." *International Journal of Applied Agriculture and Apiculture Research*, vol 8, no. 1.

Adedeji, T. A. et al. (2012) "Effect of Wattle Trait on Body Sizes and Scrotal Dimensions of Traditionally Reared West African Dwarf (WAD) Bucks in the Derived Savanna Environment." *Iranian Journal of Applied Animal Science*, vol. 2, issue 1, pp. 69–72.

Ajmone-Marsan, P. et al (2014) "The Characterization of Goat Genetic Diversity: Towards a Genomic Approach." *Small Ruminant Research*, vol. 121, pp. 58–72.

Aziz, M. A. (2010) "Present Status of the World Goat Populations and Their Productivity." *Lohmann Information*, vol. 45, pp. 42–52.

Briefer, E. et al. (2015) "Emotions in Goats: Mapping Physiological, Behavioural and Vocal Profiles." *Animal Behaviour*, vol. 99, pp. 131–63.

Briefer, E. et al. (2012) "Mother Goats Do Not Forget Their Kids' Calls." *Proceedings of the Royal Society B: Biological Sciences*, vol. 279, pp. 3749–55.

Dubeuf, J. P., and Boyazoglu, J. (2009) "An International Panorama of Goat Selection and Breeds." *Livestock Science*, vol. 120, pp. 225–31.

Ebozoje, M. O. (1998) "Colour Variation and Reproduction in the West African Dwarf (WAD) goats." *Small Ruminant Research*, vol. 27, issue 2, pp. 125–30.

Fritz Jr., W. (2017) "Self-enuriniation in the Domesticated Male Goat (*Capra hircus*)." Dissertation submitted to the Graduate School-New Brunswick, Rutgers University.

Gall, C. (1980) "Report on the Relationship between Body Conformation and Production in Dairy Goats." *The Journal of Dairy Science*, pp. 1768–81.

Heffner, R., and Heffner, H. (1990) "Hearing in Domestic pigs (*Sus scrofa*) and Goats (*Capra hircus*)." *Hearing Research*, vol. 48, pp. 231–40.

Hermiz, H. N. et al. (2016) "Inheritance of Wattle and their Effect on Twinning and Some Productive Traits in Shami Goat." *International Journal of Advances in Science Engineering and Technology*, vol. 4, issue 3.

Lenstra, J. A. et al. (2017) "Microsatellite Diversity of the Nordic Type of Goats in Relation to Breed Conservation: How Relative Is Pure Ancestry?" *Journal of Animal Breeding and Genetics*, vol. 134, pp. 78–84.

Lickliter, R. (1984) "Hiding Behavior in Domestic Goat Kids." *Applied Animal Behaviour Science*, vol. 12, issue 3, pp. 245–51.

Luikart, G. et al. (2006) "Origins and Diffusion of Domestic Goats Inferred from DNA Markers." In Zeder, M., and Bradley, D. G. (eds) *Documenting Domestication: New Genetic and Archaeological Paradigms*, pp. 294–205. Berkeley and Los Angeles, University of California Press.

Luikart, G. et al. (2001) "Multiple Maternal Origins and Weak Phylogeographic Structure in Domestic Goats." Proceedings of the National Academy of Sciences of the USA, 98(10), pp. 5927–32.

Naderi, S., Taberlet, P. et al. (2007) "Large-Scale Mitochondrial DNA Analysis of the Domestic Goat Reveals Six Haplogroups with High Diversity." *PLOS One*, published online October 10, 2007.

Naderi, S. et al. (2008) "The Goat Domestication Process Inferred from Large-scale Mitochondrial DNA Analysis of Wild and Domestic Individuals." *PNAS*, 105(46), pp.17659–64.

Pereira, F. et al. (2009) "Tracing the History of Goat Pastoralism: New Clues from Mitochondrial and Y Chromosome DNA in North Africa." *Molecular Biology and Evolution*, 26(12), pp. 2765–73.

Price, E. O. et al. (1980) "Behavioral Responses to Short-term Social Isolation in Sheep and Goats." *Applied Animal Ethology*, vol. 6, pp. 331–39.

Silanikove, N. (1997) "Why Goats Raised on Harsh Environment Perform Better than Other Domesticed Animals." *Recent Advances in Small Ruminant Nutrition*, pp. 185–94.

Skapetas, B., and Bampidis, V. (2016) "Goat Production in the World: Present Situation and Trends." *Livestock Research for Rural Development*, vol. 28, article 200.

Tuncer, S., Şireli, H., and Tatar, A. (2016) "Behavioral Patterns of Goats." Conference paper, International Scientific Agriculture Symposium, Agrosystem 2016, Jahorina, Bosnia.

Zeder, M. A. (2006) "A Critical Assessment of Markers of Initial Domestication in Goats (*Capra hircus*)." In Zeder, M., and Bradley, D. G. et al. (eds) *Documenting Domestication: New Genetic and Archaeological Paradigms*, pp. 181–208. Berkeley and Los Angeles, University of California Press.

Zeder, M., and Hesse, B. (2000) "The Initial Domestication of Goats (*Capra hircus*) in the Zagros Mountains 10,000 Years Ago." *Science*, vol. 287, pp. 2254–57.

COGNITION STUDIES

Bellegarde, L. et al. (2017) "Face-based Perception of Emotions in Dairy Goats." *Science Direct*, vol. 193, pp. 51–59.

Briefer, E. et al. (2014) "Goats Excel at Learning and Remembering a Highly Novel Cognitive Task." *Frontiers in Zoology*, vol. 11, p. 20.

Briefer, E. et al. (2013) "Rescued Goats at a Sanctuary Display Positive Mood after Former Neglect." *Applied Animal Behaviour Science*, vol. 146, pp. 45–55.

Nawroth, C. et al. (2014) "Exclusion Performance in Dwarf Goats (*Capra aegagrus hircus*) and Sheep (*Ovis orientalis aries*)." *PloS One* (journals.plos.org/plosone/article?id=10.1371/journal.pone.0093534)

Nawroth, C. et al. (2018) "Goats Prefer Positive Human Emotional Facial Expressions." Royal Society of Open Science (royalsocietypublishing.org/doi/10.1098/rsos.180491)

WEBSITES

American Consortium for Small Ruminant Parasite Control (wormx.info)

Biology of the Goat (goatbiology.com)

Breeds of Livestock—Goat Breeds Oklahoma State University (afs.okstate.edu/breeds/goats)

E (Kika) de la Garza Institute for Goat Research, Langston University, Oklahoma (luresext.edu)

Goat Resources, New South Wales Department of Primary Industries (dpi.nsw.gov.au/animals-and-livestock/goats)

Maryland Small Ruminant Page, University of Maryland (sheepandgoat.com)

Biographies

Sue Weaver (author)

Sue Weaver began writing professionally in 1969 when her first article was published in *The Western Horseman* magazine. Since then she's written hundreds of published articles, first specializing in horse magazines, then publications affiliated with *Hobby Farms* magazine. She's also written 12 books relating to livestock or poultry, including pigs, goats, sheep, cattle, chickens, donkeys, llamas, and alpacas.

Sue lives on a 29-acre property in the southern Ozarks near Mammoth Spring, Arkansas, along with her husband and a huge animal family composed of horses, full-size and miniature; a donkey; Miniature Cheviot and Katahdin sheep; Boer, Nubian, Alpine, Nigerian Dwarf, Mini LaMancha, and mixed breed goats; a llama; a pet razorback hog; and a passel of dogs, most of which are former rescues. Animals are her passion and her friends; she is a vegetarian and does not eat them.

Debbie Cherney (consultant)

Debbie Cherney (BS, animal science, 1980, University of Florida; MS, agronomy, Louisiana State University; PhD, animal nutrition, 1989, University of Florida; MA, bioethics, Medical College of Wisconsin) is a Professor of Animal Science at Cornell University. Debbie teaches three undergraduate courses: Introduction to Animal Nutrition, Introduction to Animal Welfare, and Ethics in Animal Science. Her program aims to improve the profitability of forage/livestock operations, while at the same time minimizing any negative effects of forage crops on the environment. Many of her studies have involved evaluating or improving laboratory or in vitro techniques to assess forage quality, the goal of this effort being to improve and standardize routine laboratory methodology. She has published over 150 peer-reviewed papers, chapters, and proceedings, and co-edited a popular book on forages (*Grass for Dairy Cattle*).

Acknowledgments

I'd like to thank the following good folks who contributed photos, information, and feedback to help make this book what it is: Nick Bohemia, Elodie Mandel-Briefer, Kim Depp, Eliya Elmquist, Ruth Pohl Hawkins, Sandra Hoffmann, Mika Ingerman, Ishana Ingerman, Ashley Kennedy, Jill Lane, Lisa McClear, Catherine Packham, Klaus Rudloff, Jen Schurman, Colleen Waits, Marc Warnke, John Weaver, Sibtain Zaheer, Elaine Embry, Orla Foley, Imran Zahid, Laurie Graham, Moonswine Farm (Hustle, Virgina), and Eileen and James Ray at Little Seed Farm. I'm also grateful to Tom Kitch, editorial director at Ivy Press, who diligently guided me along the way; to Sharon Dortenzio, who sourced some great photos; and to Jane Lanaway, who designed a truly stunning-looking book. And last but far from least, I'm especially grateful to Angela Koo, who is the best project editor, ever!

Index ~

A

abomasum 38, 39, 89, 107
adolescents 33
adults 33
affection 73, 114
age
  determination of 41
  life expectancy 11, 33
age of exploration 24–5
aggression 11, 80, 110–11
  butting 79, 110, 111
  feeding time 111
  fighting 78–9
  signs of 110
  agouti gene 60
alarm snort 93
Ali Kosh 18
alleles 59, 60
Alpine/French Alpine 123,
  165
  British Alpine 176
  Mini-Alpine 181
Amalthea 144
American Cashmere 206
American Pygmy 199
anatomy 30–1, 105
Angora 33, 130–1, 188
appendicular skeleton 35
Appenzell 197
Arapawa 209
archaeology 16, 20, 22, 48

Aristotle 158
axial skeleton 34–5

B

bad behavior
  see aggression
Bagot 186
bagpipes 152
Barbari 63, 178
barber pole worms 107
beards 64, 158
Becker disease 67
beer 151
Beetal 18, 33, 155
behavioral enrichment
  116–19
Bergmann, Carl 68
Bewick, Thomas 8
bezoar ibex 18, 21, 61, 164
Bilberry goats 77, 163
billy goat 8
blubbering 92
Bock beer 151
Boer 56, 57, 93, 127, 163,
  183
bonding 45, 86
bottle raising 39, 87, 103
brain 31
British Alpine 176
British Guernsey 193
British Pygmy 182

British Saanen 33, 179
British Toggenburg 196
brush control 134–5
bucklings 32, 91
bucks 8, 32, 80–3
  blubbering 92
  buck scream 92
  castration 105
  courtship ritual 83
  giving milk 57
  herd king 77
  maturity 82
  penis 80, 81, 91, 105
  smell of 11, 65, 80–1
  urinary tract 80–1, 104–5
  urolithiasis 104–5
bukkehorn 153
butting 79, 110, 111

C

Carpathian 205
cashmere 132–3, 206, 208
castration 105
cave paintings 16
cecum 39
cellulose 36, 37, 38
cheese 125
children 111
children's books 157
Christianity 147
claws 35, 46, 47

climate 74
climbing 8, 47, 117, 118,
  119
coat colors 60–3
cognition 94–5
color vision 43
colostrum 87
communication 11, 92
Cook, James 25
Corsican 49, 174
courtship ritual 83
cysts 53

D

dairy goats 11, 65, 78, 95,
  123, 137
dam 32
Damascus 33, 175
dancing goats 47, 73
dental malocclusion 41,
  163
dew claw 35, 46
deworming 107
dietary supplements 105,
  106
digestive system 36–9
  kids 39, 89
disbudding 49, 51, 90
distress calls 73, 92
DNA 20, 58
doeling 32

does 8, 32, 84–5
  freshening 55, 123
  in heat 84–5, 92
  kidding 100–3
  milk *see* milk
  wet nurses 125
dogs
  guardian dogs 109
  as predators 108–9
domestication 9, 16–21, 48
domestication syndrome
  18
dominance 73, 77, 78–9,
  111
dorcas horns 49
drums 153
drying off 55
Dutch Landrace 49, 162,
  172
dwarfism 68–9

E
ears 31, 44
esophagus 37, 39
evolution of breeds 162–3
eyes 42–3, 65
  shine in the dark 43
  see also vision

F
facial expressions 43, 94–5
fainting goats *see* myotonic
  goats
family groups 72–3
feeding behavior 40, 65,
  74–5, 111
feral goats 26–7, 32, 49,
  163
Fertile Crescent 16, 17, 20
fiber 64, 130–3
fighting 78–9

film 156
fishtail teats 57
flehman response 44, 81
flocking instinct 65, 73,
  113
folk songs 153
folklore 142–5
freshening 55, 123
friendship groups 72–3
fused teats 56–7

G
gait 47
galloping 47
Gandhi, Mahatma 155
Ganj Dareh 18, 20, 22
genetics 20, 51, 58–9
  selective breeding 18, 163
German Improved White
  33, 173
goat men 139
goatling 33
goatskin 131, 152, 153
gods 142–3, 146
Golden Guernsey 33, 170
Great Orme 27, 150
guardian dogs 109
Guernsey
  British Guernsey 193
  Golden Guernsey 33, 170
gynecomastia 57

H
hair 31, 64, 133
hand breeding 85
handling 112
haplogroups 20
Happi 155
harness goats 138–41
Harrison, Benjamin 155
head-butting 79, 110, 111

health issues
  mineral supplements 105,
    106
  nematodes 107
  selective breeding 163
  selenium deficiency 106
  urolithiasis 104–5
  white muscle disease 105
  worms 107
hearing 31, 44
herd king 77
hereditary traits 20, 58–9
heritage breeds 163
hierarchy 73, 76–9, 91
Hinduism 142, 146
hooves 35, 46–7
  caring for 98–9
  claws 35, 46, 47
  dew claw 35, 46
  overgrown 98
  trimming 98, 99
horns 11, 48–51, 64
  advantages of 50
  disbudding 49, 51, 90
  dorcas 49
  downsides to 51
  incipient corkscrew 49
  injury 51
  kids 51, 90
  polled goats 51, 59
  scimitar 48–9, 50
  scurs 49
  sheep 64
  structure 48
  types 48–9
house goats 115
human interaction 95, 114

I
Icelandic 33, 162, 171
immune system 73, 87

incipient corkscrew horns
  49
infertility 51
inheritance of traits 20,
  58–9
intelligence 94–5
intersex kids 51
intestines 38, 39
Islam 147
isolation 73

J
Josephine, Empress 132
Judaism 147

K
Kalahari Red 187
Kashmiri goats 27, 132,
  163
  see also cashmere
keratin 46, 48
kidding 100–3
  abnormal positions 102
  birth 103
  chorion 102
  first-stage labor 101
  hard labor 102–3
  normal delivery 102
  placenta 103
  positions 102
  post-delivery 103
  second-stage 102–3
  signs of 92, 100
kids 32, 86–91
  bonding 86
  bottle raising 39, 87
  bucklings 91
  colostrum 87
  digestion 39, 89
  disbudding 51, 90
  hierarchy 77, 91

horns 51, 90
intersex 51
lying out 65, 88
nursing by hand 39, 87, 103
play 91
stomach 89
suckling 39, 86–7
teeth 41
vocalization 92
Kiko 127, 184
Kinder 68, 180

L
lactation 54–7
LaMancha 123, 210
    Mini LaMancha 69, 190
lambs 89
landrace goats 49, 162, 172, 192
Lappgetter 162, 211
leather 131
Leclerc, Georges-Louis 158
leg bones 35
Leroy, Alphonse 125
life cycle 32–3
life expectancy 11, 33
lifestyle 72–5
Lincoln, Abraham 155
lips 40, 65
livestock guardian dogs 109
Llandudno 27, 150, 163
Lucy the goat 155
lying out 65, 88
Lynton goats 163

M
McCartney, Charles 139
mammals 31
mammary system 54–7

markhor 15, 21
markings 63
mascots 148–51, 156
Mastana 155
mastitis 56, 57
mating 11, 85
    courtship ritual 83
    does in heat 84–5, 92
    hand breeding 85
Mayberry, H. H. 66
meat 11, 27, 126–9
melanin 60
memory 94–5
migration 23, 24–5
military mascots 148–51
milk 11, 31
    bucks giving 57
    colostrum 87
    composition 123, 124
    dairy breeds 123
    drying off 55
    health benefits 124
    let-down 55
    mammary system 54–7
    milking through 55
    nursing by hand 103
    production 54–5, 122–4
    suckling 39, 86–7
mineral supplements 105, 106
Mini LaMancha 69, 190
Mini-Alpine 181
Mini-Nubian 195
Miniature Silky Fainting Goat 67, 69, 185
mohair 130–1
Mongolian Cashmere 208
monkey-mouthed 41
Mostyn, Sir Savage 27
mouth 37, 41, 118
Murphy 155

musical instruments 152–3
myotonic goats 47, 66–7, 185, 201
mythology 142–5

N
Nachi goats 47, 73
nanny goat 8
Nanny and Nanko 155
nematodes 107
Nigerian Dwarf 69, 82, 95, 123, 137, 194
night vision 43
Nigora 133, 191
Nirmala 155
Nubian/Anglo-Nubian 33, 93, 123, 163, 168
    Mini-Nubian 195
nursing by hand 39, 87, 103

O
Oberhasli 123, 169
Old English Goat 162, 198
Old Irish Goat 162, 207
omasum 38, 39, 89
overgrown hooves 98
Overland Jack 139

P
packgoats 136–7
pain threshold 45, 75
Pan 143
paneer 125
Pavlov, Ivan Petrovich 75
pecking order 77
penis 80, 81, 91, 105
Perry, Matthew 25
pheromones 44, 81, 114
placenta 103
play 91, 116–19

Pliny the Elder 133
polled goats 51, 59
precocious udders 57
predators 108–9
prey animals 42, 43
proverbs 159
pseudo-hermaphrodites 51
Pygmy goats 18, 68–9, 93, 137, 182, 199
Pygora 68, 133, 200

Q
quotes 158

R
raising by hand 39, 87, 103
rank see hierarchy
reasoning ability 94, 95
reticulum 38, 39
ribs 34, 35
Robinson, Phil 158
Rose, John 139
roughage 89
Royal Welsh 150–1
rumen 37, 39, 89
run impulse 75

S
Saanen/Sable 116, 123, 189
    British Saanen 33, 179
sacrificial goat 146–7
San Clemente Island 162, 202
Santeria 147
scimitar horns 48–9, 50
screaming 75, 88, 92
    buck scream 92
scurs 49
selective breeding 18, 163
selenium 106

Selkirk, Alexander 25
seniors 33
senses 42–5
Sergeant Bill 148–9
sexual maturity 32, 33
sheep 64–5, 95
  lambs 89
ships 24–5, 26
Sianis, William 155
sight 42–3
Sinovia 155
sire 32
skeleton 34–5
skull 31, 34, 35
smell
  of bucks 11, 65, 80–1
  flehmen response 44, 81
  sense of 45
sneezing 93
sow-mouthed 41
Spanish 162, 203
spinal column 34–5
split teats 57
Sparky 154
Sponenberg, Phillip 61
sport mascots 156
stomach 36, 37, 38, 89
stones (urolithiasis) 104–5
stress 73
suckling 39, 86–7
supernumerary teats 56
supplements 105, 106
Swedish Landrace 162,
  192

T

tails 35, 64
taste, sense of 45
taxonomy 14–15
teats 54, 55, 56–7

defects 56
  fishtail 57
  fused 56–7
  male goats 57
  split 57
  supernumerary 56
  see also udders
teeth 33, 40–1
  malocclusion 41, 163
tethering 22, 109
thermoregulation 50, 137
Thomsen disease 67
throwing 111
Thuringian 204
Toggenburg 123, 177
  British Toggenburg 196
touch, sense of 45
toys 118, 119
trading routes 22–3
training 112–15
  clicker training 114
  house goats 115
  packgoats 137
trampolines 117, 119
trimming hooves 98, 99
trotting 47
twisted horns 49

U

udders 54, 56, 89
  defects 55
  first kidding 100
  mastitis 56, 57
  precocious 57
  teats see teats
urinary tract 80–1
  health issues 104–5
urine-spraying 11, 75,
  80–1, 114
urolithiasis 104–5

V

Valais Blackneck 33, 50,
  113, 166
Victoria, Queen 27, 132,
  150
vintage photos 140
vision 42–3
  color vision 43
  field of vision 42, 43
  night vision 43
vocalization 44, 92–3
  alarm snort 93
  blubbering 92
  buck scream 92
  distress calls 73, 92
  individual voice 93
  kids 92
  screaming 75, 88, 92

W

walking gait 47
wattle cysts 53
wattles 52–3
weather 74
West African Dwarf 33,
  53, 56, 68, 167
wet nurses 125
wether 32
Whiskers the goat 155
white markings 63
white muscle disease
  (WMD) 105
white spotting 60, 63
William de Goat 149
witch's milk 57
worms 107

Y

yearling 33
Yule goat 145

Picture Credits

The publisher would like to thank the following for permission to reproduce copyright material:

Alamy/AF archive: 154T; Artokoloro Quint Lox Limited: 144B; Brian Harris: 74; Chris Howes/Wild Places Photography: 65B; chrisstockphoto: 153T; Chronicle: 146B, 148C; Cosmo Condina Western Europe: 112; Dawna Moore: 45; Ernie Janes: 53T; Everett Collection Historical: 155T, 156B; Florilegius: 21; FLPA: 27T; Forget Patrick/Sagaphoto.com: 123; Ger Bosma: 172; Gloria Good: 93; Granger Historical Picture Archive: 25T; Iakov Filimonov: 17BR; Ida Baranyai: 19; Idealink Photography: 16; ImageBroker: 53B; IndiaPicture: 147; Inga Spence: 23BL; Interfoto: 22T, 143L, 146T; Jean-Yves Roure: 197; Julie Shipman: 133B; Lebrecht Music & Arts: 152L; Robert Melen: 150B; Some Wonderful Old Things: 46T; SOTK2011: 151B; The Book Worm: 40T; The Picture Art Collection: 34, 142; Trinity Mirror/Mirrorpix: 150T; UPI: 154L; Vladislav Vladimirov: 92R; Wildlife GmbH: 173, 204.

Elodie F. Briefer: 94B.

Kim Depp (InMotionPhotos.biz): 68T, 133T, 200.

Dreamstime/Alexpolo: 15BL (Capra caucasica), Leonid Andronov: 175, Robin M Coventry: 184.

Eliya Elmquist (Green Gables MiniNubians): 195.

Elaine Embry (Kickapoo Creek Nigerians): 115.

Getty Images/Anadolu Agency/Contributor: 96, Archive Photos/Transcendental Graphics/Contributor: 139, Bongarts/Anja Heinemann/Staff: 156T, DEA Picture Library/Contributor: 24, E+/wanderluster: 113, Hulton Archive/Heritage Images/Contributor: 157T, Hulton Archive/Imagno/Contributor: 155B, Hulton Archive/Print Collector/Contributor: 144T, Hulton Fine Art Collection/Heritage Images/Contributor: 132T, iStock/canaran: 86, iStock/edmdusty: 75T, iStock/flyingrussian: 87B, iStock/mlharing: 171, Lonely Planet Images/John Elk III: 174, Moment/Elizabeth W. Kearley: 118B, The LIFE Picture Collection/Time Life Pictures/Contributor: 157B, ullstein bild/Contributor: 22B, Universal Images Group/Christophel Fine Art/Contributor: 159B, Universal Images Group/VW Pics/Contributor: 105T.

Laurie Graham, courtesy of Moonswine Farm, Hustle, VA, USA: 66.

M. Hickey: 77.

Imperial War Museum/Richard Nash (CH 8999): 149.

Mika/Ishana Ingerman: 141T.

Ashley Kennedy: 180.

Jill Lane: 117B, 169.

Delana Lefevers: 202.

Library of Congress Prints and Photographs Division: 130BL, 138B, 140B, 140C, 140T, 148B, 151T.

Lisa McClear: 181.

Klaus Rudloff: 23TL, 192.

Shutterstock: 1, 2, 3TR, 3B, 4B, 5T, 5B, 6, 11, 13, 14M (Artiodactyla), 14M (Caprinae), 14M (Bovidae), 14T, 14C (Capra), 15 (all except Capra caucasica), 17TR, 17TCR, 17TL, 17TCL, 17TCL, 18, 20, 23BR, 26T, 26B, 27B, 28, 31, 32T, 32BL, 32BCR, 32BCL, 37, 38, 39T, 39B, 40B, 42T, 42BR, 44 (Graph, CT, T, B), 48T , 50B, 51L, 51R, 52T, 54T, 55, 57T, 57B, 58L, 59R, 59L, 60T, 60B, 64T, 64CT, 67, 71, 72T, 72B, 73, 76B, 78, 79, 82T, 82B, 83T, 84BL, 84BR, 87T, 88B, 88T, 89T, 91R, 91L, 92L, 94T, 98B, 98T, 98C, 100, 104T, 105B, 106TR, 108, 109B, 109TR, 110T, 110B, 111B, 111T, 114T, 118B, 119, 122B, 122T, 124, 125T, 126B, 126T, 127, 128, 129T, 129B, 130BR, 131TL, 131TR, 131B, 131TC, 132B, 134T, 134B, 136, 138T, 143R, 148T, 152R, 153B, 154R, 158T, 159T, 162B, 163L, 164, 166, 182, 187, 198, 205, 208, 209, 211, 218, 219.

Unsplash/Tristan Gevaux: 7; Anoop Bhaskar: 120; mana5280: 8; Arjan Stalpers: 9; Agnieszka Kowalczyk: 47; Thierry Chabot: 104B; Sergio Souza: 154B; Simon Matzinger: 160.

Colleen Wait: 95.

Marc Warnke (PackGoats): 10, 137T, 137B.

Sue Weaver: 4T, 43T, 49B, 52B, 56B, 65C, 65T, 68C, 69T, 81R, 83B, 85, 90BL, 90BC, 90BR, 90T, 101CL, 101TL, 101TR, 101BR, 101CR, 101BL, 102BC, 102BR, 102BL, 103BL, 103BC, 103BR, 103T, 106TL, 107B, 109TL, 114B, 116, 117T, 165, 167, 210.

Wellcome Collection (CC BY 4.0): 25B, 76T, 125B, 213.

Wikimedia Commons/Museum of Veterinary Anatomy FMVZ USP/Wagner Souza e Silva (CC BY-SA 4.0): 35T; Helena Bowen and Richard Bowen (CC BY-SA 3.0): 41T; Jenny Nystrom: 145B; Tony Nordin (CC BY-SA 3.0): 145T.

Sibtain Zaheer: 63B.

All reasonable efforts have been made to trace copyright holders and to obtain their permission for the use of copyright material. The publisher apologizes for any errors or omissions and will gratefully incorporate any corrections in future reprints if notified.